MELAB

Computer Programs with Software
for Mechanical Engineers

ANDREW D. DIMAROGONAS

W. Palm Professor of Mechanical Design
Washington University, St. Louis, Missouri

PRENTICE HALL, Englewood Cliffs, New Jersey 07632

Production Editor: Alan Dalgleish
Acquisitions Editor: Doug Humphrey
Supplements Acquisitions Editor: Alice Dworkin
Prepress Buyer: Linda Behrens
Manufacturing Buyer: Dave Dickey

Printed in the United States of America

10 9 8 7 6 5 4 3 2 1

ISBN 0-13-951757-X

Prentice-Hall International (UK) Limited, *London*
Prentice-Hall of Australia Pty. Limited, *Sydney*
Prentice-Hall Canada Inc., *Toronto*
Prentice-Hall Hispanoamericana, S.A., *Mexico*
Prentice-Hall of India Private Limited, *New Delhi*
Prentice-Hall of Japan, Inc., *Tokyo*
Simon & Schuster Asia Pte. Ltd., *Singapore*
Editora Prentice-Hall do Brasil, Ltda., *Rio de Janeiro*

To Dr. Hans F. Bueckner

Distinguished Mathematician and Man of Principles

who taught me that

WE SHOULD NOT EXPECT THE COMPUTER

TO DO ALL THE WORK

CONTENTS

Contents

PREFACE

The general availability of microcomputers and computer methods of instruction at the undergraduate level and the increasing need for ABET demanded design content, is prompting undergraduate courses in Mechanical Engineering to focus on design applications and open ended problems. Many times this requires extensive computations which can be greatly facilitated with computer utilization. This prompted the author to organize a set of computer programs to assist undergraduate instruction in mechanical engineering.

Most of the underlining theory is included in the following textbooks:

a. Dimarogonas, A.D., Computer Aided Machine Design, Prentice Hall, 1988.

b. Dimarogonas, A.D., Haddad, S.D, Vibration for Engineers, Prentice Hall 1992.

c. Nikravesh, P., Computer Aided Analysis of Mechanical Systems, Prentice Hall 1988.

The general switch from batch processing to interactive mode and extensive use of graphics made necessary the deviation from the simple routines printed in textbooks to more organized programs with pre- and post- processing capabilities and screen graphics.

Tantamount to the interactive use is the introduction of the SIMULAB idea, the simulation of laboratory experiments on the computer screen where the student runs a "real" experiment and obtains simulated experimental results. Even the scatter of results are built-in in the form of random disturbances of the operating parameters. Typical example is the Balancing Machine Simulator. There is an initial unbalance on a two-disk rotor, random or prescribed by the instructor. The student operates the machine from the keyboard and has a real time view of it on the screen. He or she performs the balancing and all the balance runs are stored in a file for evaluation by the instructor. They see also the results of balancing on the instruments simulated on the screen. Even the sound of the machine is simulated.

Chapter 1 introduces the user to the **MELAB** operation and to the COMPUME menu.

Chapter 2 contains basic numerical methods which will be used later. This chapter is not intended only to support a course in numerical analysis but also to support the numerical needs of later chapters.

Chapter 3 contains basic graphics programs used later, mostly for input preparation of other programs.

Chapter 4 describes modules for structural analysis, using transfer matrix or finite element techniques.

Chapter 5 introduces computer programs for kinematic and dynamic analysis of planar mechanisms.

Chapter 6 introduces SIMULAB, a simulated vibration laboratory, and vibration analysis and signal processing programs.

Chapter 7 includes programs for machine dynamic analysis. An educational edition of RODYNA, a general purpose Rotor Dynamic analysis program, is presented.

Chapter 8 describes programs used for diagnosis and prognosis and the educational edition of EXPERTS, a Neural Network expert system shell.

Chapter 9 includes material failure analysis, used in Strength of Materials and Machine Design courses.

Chapter 10 presents an optimization program and other design aids, for Machine Design courses.

Chapter 11 presents computer programs for design of fasteners, screws, rivets, welds, used in Machine Design courses.

Chapter 12 presents modules for rotary drives design, as used in Machine Design courses

Chapter 13 is devoted to finite element solution of the bearing lubrication equation.

For every program in this package, the following material is included here:

a. A summary of the basic theory, program description and references for further study.

b. Program documentation and user instructions.

c. Executable or linlable code on floppy disks.

Though most of the programs have been developed for the author's textbooks, this volume could be used with other mechanical engineering textbooks. **MELAB** is a organized set of computer programs designed for mechanical engineers and students of mechanical engineering. They are designed to support the study of courses primarily in mechanical design and vibration but they can be also used with courses in computer aided design, finite elements, numerical methods. They are not to substitute for special purpose commercial computer programs, for example drafting (AUTOCAD, CADKEY, etc), structural analysis (NASTRAN, ANSYS, etc), simulation (ADAMS, DRAM, TUTSIM, DADDS, etc), etc., since they have a different scope and a broader range.

Most of the programs are in executable code. Programs which require the user to write code, OPTIMUM or POINCARE, for example, are in MICROSOFT-linker compatible object code, so that the user supplied part can be written in MICROSOFT Basic, C or FORTRAN. For most of the programs, except for RODYNA and EXPERTS, the source code is available for the instructors, to be used for development and programming assignments. These programs are for educational use, therefore the size of the problems is limited to the one required to demonstrate the principles involved.

Your comments, suggestions, criticism and contributions for future editions will help making this package better.

Andrew D. Dimarogonas

Mechanical Engineering, Campus Box 1185,
Washington University, 1 Brookings Dr.
St. Louis, Missouri 63130
e-mail: ADD@ MECF.WUSTL.EDU
FAX: (314) 935-4434

While this book was in the manufacturing process, Math Works, Inc., introduced SIMULAB (TM), Block Diagram Software for Nonlinear Simulation of Dynamic Systems. It should be noted that the name SIMULAB in this book is used as a shorthand name for "SIMUlated vibration LABoratory", a group of computer programs which simulate vibration experiments hardware, and it is not related in any way to nor it is a substitute for SIMULAB, which is a registered trade mark of Math Works, Inc.

1

CHAPTER ONE
INTRODUCTION

The computer programs presented in this package cover a wide area of topics and emphasize the fact that computer aided design is not merely graphics or even finite element analysis but is a much wider topic.

Before the user starts using the programs, he or she must become familiar first with the hardware and with the particular problem at hand. In particular, he or she must be familiar with a) the method of analysis used and b) with the numerical procedure used.

No computer program can be better than the analytical and numerical method it utilizes. Moreover, some problems have empirical constants or equations in their solution and the user must be able to make this distinction, in particular when compares results of the code with results obtained by hand or by another computer program on the same problem. For example, when solving a beam problem, the results of the finite element or the transfer matrix method should be essentially the same, differing only due to the numerical error which might me different in the two methods. On the other hand, for the design of gears, a number of empirical equations is used and the results of two solutions should be essentially the same only if the same equations and the same constants are used. Many times this is not the case but the results should not be drastically different.

There is no general method to test the numerical accuracy of the results of numerical computations. There are some hints, however, which can be used to enhance the confidence of the user in the results:

1. The continuity test. Variation of an input quantity by a small amount should result in the same order of variation of the result quantities. For example, in a beam deflection problem, variation of one load by 2%, 4%, 6% should result in similar variation of the results, ie, 2%, 4%, 6%, etc. In some problems, such as in gears where the modul has standard values, the above behavior is expected to have a stepped character. Wild variations of the output for smooth variation of the input indicates numerical problems.

2. The interval halving test. Problems involving a user defined step, such as finite element analysis of a straight beam-length of each element, integration of differential equations-time step, if the step is changed to 1/2 the variation of the results should be minimal. In fact, for some problems this variation yields an estimate of the numerical error. There is a limit for the extent that this can be done because at very small steps the round-off error might be substantial.

Finally, the reader must read the README.TXT file before using the program for the latest information and updates.

1.1. Setting-up the hardware.

MELAB is a set of computer programs to support mechanical engineering courses, such as Design, Vibration, Numerical Analysis. This is a DOS version and can run in any DOS-compatible machine.

To run **MELAB**, you need the following:

An IBM AT 286 or 386 or 486 compatible computer or any machine which runs DOS 3.2 or later, or OS/2 or UNIX simulation of DOS.

A CGA, EGA, VGA, HERCULES or EGA-Monochrome card. Some modules require EGA card.

A color monitor. A monochrome graphics monitor with high resolution, 640x350, is sufficient for most programs.

A mathematics co-processor can speed-up execution considerably but it is not essential.

The package should be used through a hard disk. Specific modules can be used with a floppy disk drive.

Screens can be captured for printing with CAPTURE* or similar software or using the <Print Scrn> key. If used with WINDOWS, <Print Scrn> will place the screen on the clipboard and can be subsequently pasted in other application programs, like WORD.

The input for most programs is entered through a text file [PROGRAM NAME].DAT and by user interaction.

The output for most programs is shown on graphics screens and it is written, for most modules, to a text file [PROGRAM NAME].RES.

1.2. Setting-up the software

a. If you want to use a hard disk drive.

You need approximately 3 Mb of hard disk space to install **MELAB**. On the **MELAB** diskette #1, there is an installation program **SETMELAB**. Put diskette #1 in drive A: and type:

A:\>SETMELAB

The installation program will create a subdirectory C:\MELAB by default or any other that the user will specify. If the programs are stored on the program diskettes in compressed form (i.e. *.ARC archives), the installation program will decompress them and save them into the specified subdirectory. If the programs are stored in executable form (*.EXE for the executable files, for example) the installation program will simply transfer them into the specified subdirectory. The user should just follow the program instructions.

If you want to use only a number of modules, to save hard disk space, you should keep in the subdirectory:

a. The COMPUME.EXE, the COMPUME.SYS and MELAB.BAT files.

b. The, SSMENU.EXE, SSMENU.TXT, SSMENU.HLP files, to use the spreadsheet input screen.

---

* A Trademark of the MICROSOFT Corporation.

c. The PULMENU.EXE and the PULMENU.TXT files, if you want to use pull-down menus.

d. The .EXE file(s) for the modules you want to include. The number of modules depends on the capacity of your diskette. Some modules need input, output or other files, listed with the description of each module. Such files should also be present in the **MELAB** subdirectory.

e. The CINSTALL.EXE file.

You are now ready to start **MELAB**. You can operate it from **WINDOWS** by loading MELAB.BAT as a non-WINDOWS application or from the DOS prompt.

To operate from DOS, load first any memory resident programs you might want to use to capture screens, such as Microsoft CAPTURE, for example. From the **MELAB** subdirectory, type

C:\MELAB> MELAB

The root menu appears on the screen. You are ready now to use **MELAB**

b. If you want to use a floppy disk drive.

You have to install the system on a hard disk first. Then, copy on the floppy diskette from which you will operate:

a. The COMPUME.EXE, MELAB.BAT and the COMPUME.SYS files of **MELAB**.

b. The, SSMENU.* files, if you want to use the spreadsheet input screen.

c. The PULMENU.EXE and the PULMENU.TXT files, if you want to use pull-down menus.

d. The .EXE file(s) for the modules you want to include. The number of modules depends on the capacity of your diskette. Some modules need input, output or other files, listed with the description of each module. Such files should also be copied on the floppy diskette.

e. The CINSTALL.EXE file.

You are now ready to start **MELAB**. Load first any memory resident programs you might want to use to capture screens, such as Microsoft CAPTURE, for example.

Place your diskette on the floppy disk drive and transfer control to this drive, i.e.[**]

C\> A:

Then type:

A\> COMPUME

The root menu appears on the screen. You are ready now to use **MELAB**

1.3. Using MELAB

a) Units.

You can use any system of compatible units for the data. The results will be in units compatible with the input data. In certain cases, such as in antifriction bearing selection, specific units are used. In such cases, the program specifies the units of input-output. In general, the SI system is suggested.

** Underlined text is typed by the user. At the end of the underlined text the user should hit the ENTER key.

b) Entering Data.

If data are entered after a ? prompt, the value is typed and the ENTER key is pressed. In most program inputs, before the ? prompt there is a default value between the <...> markers which is built into the program or it is the value entered already by the user previously. If you type any text after the ? prompt and hit ENTER, the default value is erased and the typed value is in the program from then on. If you want to maintain the default value, just hit ENTER. If you are prompted to enter a command, usually a single letter, without the ? prompt, just hit the appropriate letter without hitting ENTER. If you are prompted to answer with a Yes or No, just enter Y or N.

c) Problem Size.

These programs are for educational use and the size of the problems is limited to the one required to demonstrate the principles involved. There is a professional version of **MELAB** available with full capabilities. In that version, the problem size is limited only by the computer expanded memory >640 kb.

d) Updating.

Later versions usually contain all features of previous editions. Input-output, however, might be slightly different. Updates are included in the README.TXT file.

e) Hard copy of the screen.

MELAB itself does not include a screen saving facility. Printed results can be obtained in the following ways:

I. Most programs write the results on an ASCII file [PROGRAM NAME].RES which can be retrieved and printed with any editor or word processor.

II. Screens can be captured with memory resident programs, mentioned above, or by the <Print Scrn> key, if DOS 5.0 is used and the system is set for printing EGA screens, as described in the DOS 5.0 manual.

III. If you use **MELAB** from WINDOWS, <Print Scrn> captures the screen and puts it in the clipboard. The screen can be subsequently pasted in, edited and saved with PAINTBRUSH and then used in application programs, such as WORD FOR WINDOWS.

f) Error Messages.

Each module has built-in error messages and some remedies, where practicable. In any case, an error code number is given. A summary of error codes is given in appendix II.

g) Using WINDOWS* .

MELAB can be used with WINDOWS. To this end, from the program manager, select Non-Windows applications. Then, in the Program Manager menu select File, New. In the description field write MELAB and in the command field write path\MELAB.BAT, where path\ is usually C:\MELAB\. The **MELAB** icon will then appear in the non-WINDOWS applications from where it can be selected. You can change the icon or move it to another window. More details are given in your WINDOWS manual.

---

* Trademark of Microsoft Corp.

1.4. COMPUME

Scope: A menu program from which all programs of the package can be initiated through user interaction.

Method: Spreadsheet selection.

Files needed: COMPUME.EXE, CINSTALL.EXE, COMPUME.SYS, SSMENU.EXE, SSMENU.TXT, SSMENU.HLP, PULMENU.EXE and PULMENU.TXT.

Use: If the program is used for the first time, the user has to inform the program what hardware he is using. To do this, you type

C:\MELAB\CINSTALL

and answer the program questions subsequently.

a. The program asks for the path for reading input files and writing output files, if other than the current directory. This is the case, for example, when the program is used from a protected directory and the user space is another directory or drive. For example, if the programs are in subdirectory C:\MELAB and the user wants to use drive A: for input-output, the response is:

Input-Output path <> : A:

Otherwise, if the input-output files are residing in the current directory, just hit ENTER or enter just a space if there is already a path that you donot want.

All input-output files with other than .EXE extensions should be in the path. All .EXE files should be in the current directory.

b. The graphics card. Select CGA, EGA, VGA, MONO, HERCULES, etc, depending on the card you have. Even for VGA systems, EGA is suggested because VGA requires a lot of memory.

c. The monitor. Select C (color), B (black and white).

d. The menu type. An S for spreadsheet menu or P for pull-down menus, should be selected.

When the configuration session is terminated, the COMPUME screen appears, in spreadsheet or pull-down menus form, as selected in CINSTALL. The configuration program maintains the file COMPUME.SYS where the configuration information is stored.

To use the program at any subsequent time type:

C:\MELAB> MELAB

A spreadsheet or pull-down menu appears on the screen with the program names. Using the keypad direction keys you highlight a program. Hitting the ENTER key runs the program. When execution is terminated, control returns to the menu program.

Advanced users can modify the contents of the spreadsheet or the pul-down menus. This can be done by editing the files SSMENU.TXT and SSMENU.HLP, or PULMENU.TXT, respectively.

The file SSMENU.TXT contains the names of the modules that the menu program runs:

```
14,5
QUIT       DESIGNDB   CINSTALL                          Utilities
SOLID      AUTOMESH                                     Graphics
BEAMSTAT   PREFRAME   FINFRAME   FINSTRES   TMCOLUMN    Structures
```

SIMUMECH				Mechanisms
SIMULAB	RODYNA	MULTILIN	POINCARE	Vibration
EXPERTS	ANALYZER			Diagnosis
ENGINE				IC Engines
HYDNET				Hydraulics
SAFFAC	FAILURE	MATERIAL	SECTIONS	Materials
OPTIMUM	FITS	BRAKE	COUPLING	Design
RIVETS	WELDS	HELICSPR	BUCKLING	Fasteners
GEARPLOT	GEARDES	WORMDES	FLATBELT	VBELTS	Drives I
ROTORDYN	SHRINKFT	SIMUDRV	SHAFTDES	SLIDERBR	Drives II
LINEQ	MATINVER	JACOBI	CHOLESKY	STODOLA	Numerical Methods

The first line has then number of rows Nr, number of columns Nc of the menu. Each subsequent line contains program names in the first 10xNc fields and then a message. Each field occupies 10 spaces. The user can:

a. Add in the empty fields names of programs with extension *.EXE (without the extension) in his current drive. The programs can be selected and run from COMPUME.

b. Replace one or more (or all) the program names with others. If a line is not needed at all, it should be left blank. There should be Nr lines in the file.

File COMPUME.HLP contains the help lines which appear on the bottom of the COMPUME screen explaining the function of a highlighted program:

```
"MELAB: COMPUTER PROGRAMS FOR MECHANICAL ENGINEERS"
14,5
1, 1, "Quits COMPUME and returns to DOS."
1, 2, "Creates and maintains data files."
1, 3, "Configures system for available hardware"
1, 4, "Reserved"
1, 5, "Reserved"
2, 1, "Solid modeling program."
2, 2, "Automatic mesh generation program. Plane triangular and
quadrilateral elements are generated. Meshes compatible with FINSTRES and
SLIDERBR."
2, 3, "Reserved"
2, 4, "Reserved"
2, 5, "Reserved"
3, 1, "Static analysis of continuous beams. The transfer matrix method is
used."
3, 2, "Preprocessing program for FINFRAME. Development of 3-D frame
models."
3, 3, "Static analysis of 3-D frames. The Finite Element Method is used.
Input should    be prepared with PREFRAME or with a text file editor."
3, 4, "Finite Element static analysis of plane problems. Input should be
prepared with    AUTOMESH or by a text file editor."
3, 5, "Stability analysis of continuous beam-columns. The transfer matrix
method is used."
```

4, 1, "Kinematic and dynamic simulation of mechanisms"
4, 2, "Reserved"
4, 3, "Reserved"
4, 4, "Reserved"
4, 5, "Reserved"
5, 1, "Simulated vibration laboratory"
5, 2, "Dynamic analysis of rotor-bearing systems"
5, 3, "Dynamics of linear, multidegree-of-freedom systems"
5, 4, "Phase Portrait and Poincare' Map of dynamic systems."
5, 5, "Reserved"
6, 1, "Generator of static neural network expert systems"
6, 2, "Signal Analysis for diagnosis and prognosis"
6, 3, "Reserved"
6, 4, "Reserved"
6, 5, "Reserved"
7, 1, "Simulation of a single cylinder IC engine. Available through the author"
7, 2, "Reserved"
7, 3, "Reserved"
7, 4, "Reserved"
7, 5, "Reserved"
8, 1, "Analysis of hydraulic networks. Available through the author"
8, 2, "Reserved"
8, 3, "Reserved"
8, 4, "Reserved"
8, 5, "Reserved"
9, 1, "Safety factor estimation with reliability analysis."
9, 2, "Failure analysis with combined loads."
9, 3, "Material physical and mechanical properties database."
9, 4, "Properties of sections. Section drawing should be made with SOLID or an appropriate data file should be prepared with an editor."
9, 5)   "Reserved"
10, 1, "Constrained optimization of a multi-variable function. The source code should be used with a high level language compiler. Not selectable from this menu."
10, 2, "Computation of limit dimensions for fits. Basic Hole system"
10, 3, "Automatic design and AUTOCAD drawings of drum brakes. Available through the author"
10, 4, "Automatic design and AUTOCAD drawings of air clutches. Available through the author"
10, 5, "Reserved"
11, 1, "Analysis of compound rivetings"
11, 2, "Analysis of compound welds"
11, 3, "Design of helical springs"
11, 4, "Design of columns of circular cross-section in compression"
11, 5, "Reserved"
12, 1, "Plotting and animation of a pair of spur gears"
12, 2, "Design of spur, helical and bevel gears"
12, 3, "Design of worm-gear drives"
12, 4, "Design of flat belt drives"
12, 5, "Design of V-belt drives"
13, 1, "Critical speeds of rotating shaft with the transfer matrix method"
13, 2, "Design of shrink fits with compound hubs"
13, 3, "Simulation of two-rotor drives with flexible coupling"
13, 4, "Design of shaft and rolling element bearing systems"
13, 5, "Design analysis of hydrodynamic slider bearings"

```
14, 1, "Solution of a system of linear equations with Gauss elimination.
Determinant evaluation."
14, 2, "Inversion of a real square matrix with Gauss elimination"
14, 3, "Eigenvalues and eigenvectors of a real square matrix with the
Jacobi method"
14, 4, "General Eigenvalue problem lamda[M] = [K] with the Cholesky
method"
14, 5, "General Eigenvalue problem lamda[M] = [K] with the Stodola
method"
```

The first line has the **MELAB** menu title.

The second line has the number of rows Nr, number of columns Nc of the menu, in this case 14,5.

The subsequent lines contain the line number, column number of the program in the menu (or the COMPUME.TXT file), the message. This message can be up-to 232 characters long (it is printed in 3 lines). There should be Nr x Nc lines. The user can modify the text in the lines if changes in the COMPUME.TXT files were made.

The PULMENU.TXT file contains both the menu and help information in a different arrangement: The first line has the number of the menus in the top horizontal menu line. Each subsequent line has the number of submenus and their names, up to 10 characters each.

The next line has the names of the horizontal menu line, up-to 6 characters each.

Each of the subsequent lines, has the horizontal menu number, the vertical menu number, the help message-up to 250 characters long, arranged consequtively, as shown-not mixed. The file follows:

```
10
5,QUIT,DESIGNDB,CINSTALL,SOLID,AUTOMESH
5,BEAMSTAT,PREFRAME,FINFRAME,FINSTRES,TMCOLUMN
6,SIMUMECH,SIMULAB,RODYNA,MULTILIN,POINCARE,CONTINUA
2,EXPERTS,ANALYZER
2,ENGINE,HYDNET
5,SAFFAC,FAILURE,MATERIAL,SECTIONS,OPTIMUM
7,RIVETS,WELDS,HELICSPR,BUCKLING,FITS,BRAKE,CLUTCH
5,GEARPLOT,GEARDES,WORMDES,FLATBELT,VBELTS
5,ROTORDYN,SHRINKFT,SIMUDRV,SHAFTDES,SLIDERBR
5,LINEQ,MATINVER,JACOBI,CHOLESKY,STODOLA
"Util","Struct","Vibr","Diagno","HeatFl","Mater","Fasten","Gears","Rotors
","NumAnl"
1,1,"Quits COMPUME and returns to DOS."
1,2,"Creates and maintains data files."
1,3,"Configures system for available hardware."
1,4,"Solid modeling program."
1,5,"Automatic mesh generation program. Plane triangular and
quadrilateral elements are generated. Meshes compatible with FINSTRES and
SLIDERBR."
2,1,"Static analysis of continuous beams. The transfer matrix method is
used."
2,2,"Preprosessing program for FINFRAME. Development of 3-D frame
models."
2,3,"Static analysis of 3-D frames. The Finite Element Method is used.
Input should   be prepared with PREFRAME or with a text file editor."
```

2,4,"Finite Element static analysis of plane problems. Input should be prepared with   AUTOMESH or by a text file editor."
2,5,"Stability analysis of continuous beam-columns. The transfer matrix method is   used."
3,1,"Kinematic and Dynamic simulation of mechanisms"
3,2,"Simulated Vibration Laboratory"
3,3,"Dynamic analysis of rotor-bearing systems"
3,4,"Dynamics of linear,multidegree-of-freedom systems"
3,5,"Phase Portrait and Poincare' Map of dynamic systems."
3,6,"Vibration analysis of continuous systems, strings, beams, membranes, plates."
4,1,"Generator of static neural network expert systems"
4,2,"Signal Analysis for diagnosis and prognosis"
5,1,"Simulation of a single cylinder IC Engine"
5,2,"Simulation of a general Hydraulic Network"
6,1,"Safety factor estimation with reliability analysis."
6,2,"Failure analysis with combined loads."
6,3,"Material physical and mechanical properties database."
6,4,"Properties of sections. Section drawing should be made with SOLID or an   appropriate data file should be prepared with an editor."
6,5,"Constrained optimization of a multi-variable function. The source code should be used with a high level language compiler. Not selectable from this menu."
7,1,"Analysis of compound rivetings"
7,2,"Analysis of compound welds"
7,3,"Design of helical springs"
7,4,"Design of columns of circular cross-section in compression"
7,5,"Computation of limit dimensions for fits. Basic Hole system"
7,6,"Automated design of drum brakes. Produces .DXF drawings for AUTOCAD"
7,7,"Automated design of inflated rubber-tube clutches. Produces .DXF drawings for AUTOCAD"
8,1,"Plotting and animation a pair of spur gears"
8,2,"Design of spur,helical and bevel gears"
8,3,"Design of worm-gear drives"
8,4,"Design of flat belt drives"
8,5,"Design of V-belt drives"
9,1,"Critical speeds of rotating shaft with the transfer matrix method"
9,2,"Design of shrink fits with compound hubs"
9,3,"Simulation of two-rotor drives with flexible coupling"
9,4,"Design of shaft and rolling element bearing systems"
9,5,"Design analysis of hydrodynamic slider bearings"
10,1,"Solution of a system of linear equations with Gauss elimination and evaluation   of determinants"
10,2,"Inversion of a real square matrix with Gauss elimination"
10,3,"Eigenvalues and eigenvectors of a real square matrix with the Jacobi method"
10,4,"General Eigenvalue problem lamda[M] = [K] with the Cholesky method"
10,5,"General Eigenvalue problem lamda[M] = [K] with the Stodola method"

help:
This file contains the definition of the pull-down menus.
Line 1: n (The number of the pull-down menus, maximum 10)
line 2: m1,word1(1),word2(1),...,wordm1(1) (number of elements of submenu 1, names of submenu 1 entries)
line 3: m2,word1(2),word2(2),...,wordm2(2) (number of elements of submenu 2, names of submenu 2 entries)
....

```
line n+1: mn,word1(n),word2(n),...,wordm1(n) (number of elements of
submenu n, names of submenu n entries)
line n+2: menu1, menu2,...,menu (menu-line entries, maximum 10, maximum 7
characters each)
```

the remaining lines contain help lines (max 250 characters each). Each
line has the form:
```
   i,j,help$(i,j)
   where i the number of submenu
         j the entry j of submenu i
         help$(i,j) the help line
ie
1,1,help$(1,1)
1,2,help$(1,2)
...
1,m1,help$(1,m1)
2,1,help$(2,1)
.....
```

Each sub-menu can have up-to 10 entries.

Probable errors: File not found. Files needed, as above, not in the proper directory. Selected file not in current directory.

COMPUTER PROGRAMS FOR MECHANICAL ENGINEERS

Chapter	MENU				Subject
1	QUIT	DESIGNDB	CONFIGUR			Utilities
2	SOLID	AUTOMESH				Graphics
3	BEAMSTAT	PREFRAME	FINFRAME	FINSTRES	THCOLUMN	Structures
4	SIMUMECH					Mechanisms
5	SIMULAB	RODYNA	MULTILIN	POINCARE		Vibration
6	EXPERTS	ANALYZER				Diagnosis
7	>ENGINE					IC Engines
8	HYDNET					Hydraulics
9	SAFFAC	COMLOAD	MATERIAL	SECTIONS		Materials
10	OPTIMUM	FITS				Design
11	RIVETS	WELDS	HELICSPR	COLUMN		Fasteners
12	GEARPLOT	GEARDES	WORMDES	FLATBELT	VBELTS	Drives I
13	ROTORDYN	SHRINKFT	SIMUDRV	SHAFTDES	SLIDERBR	Drives II
14	LINEQ	MATINVER	JACOBI	CHOLESKY	STODOLA	Numerical Methods

Not available in this version

Locate program with keypad arrows and hit ENTER

2

CHAPTER TWO
BASIC NUMERICAL METHODS

2.1. Systems of LINear EQuations

Function: Solution of a system of linear equations with the method of Gauss Elimination.

References: Faddeeva, V.N., 1959, Meirovitch, L., 1980, Dimarogonas 1988.

Hardware Requirements: 512k, 1 FD 360k, any monitor card, monitor.

Limitations: Up to 20 equations.

Files needed: LINEQ.EXE, GAUSSEL.EXE, LINEQ.DAT

Method: Assume a system of linear equations

$$
\begin{aligned}
a_{11}x_1 + a_{12}x_2 + \cdots + a_{1n}x_n &= b_1 \\
a_{21}x_1 + a_{22}x_2 + \cdots + a_{2n}x_n &= b_2 \\
&\cdots\cdots\cdots\cdots\cdots\cdots\cdots\cdots\cdots\cdots\cdots \\
a_{n1}x_1 + a_{n2}x_2 + \cdots + a_{nn}x_n &= b_n
\end{aligned}
\qquad (a)
$$

In matrix form

$$\mathbf{Ax = b}$$

a. Real matrix A and vector b

Multiply the equation i, i=2,3,..,n, by $-a_{11}/a_{i1}$ and add to it the first equation. The resulting system is:

$$
\begin{aligned}
a_{11}x_1 + a_{12}x_2 + \cdots + a_{1n}x_n &= b_1 \\
a_{22}x_2 + \cdots + a_{2n}x_n &= b_2 \\
\cdots\cdots\cdots\cdots\cdots\cdots\cdots\cdots\cdots\cdots\cdots & \qquad (b) \\
a_{n2}x_2 + \cdots + a_{nn}x_n &= b
\end{aligned}
$$

where $a_{ij} =: a_{ij} - a_{ij}a_{11}/a_{i1}$, $b_i =: b_i - b_i a_{11}/a_{i1}$ and the symbol =: means substitution.

The system (a) is equivalent to the equation,

$$a_{11}x_1 + a_{12}x_2 + \cdots + a_{1n}x_n = b_1$$

and the system of dimension (n-1)(n-1)

$$a_{22}x_2 + \ldots + a_{2n}x_n = b_2$$
$$\ldots \ldots \ldots \ldots \ldots \ldots \ldots \ldots \ldots \ldots$$
$$a_{n2}x_2 + \ldots + a_{nn}x_n = b_n$$

(c)

The system (c) can be further reduced to the its first equation and a system of dimensions (n-2)(n-2). This process is repeated until the nth equation and the system then becomes

$$a_{11}x_1 + a_{12}x_2 + \ldots + a_{1n}x_n = b_1$$
$$a_{22}x_2 + \ldots + a_{2n}x_n = b_2$$
$$\ldots \ldots \ldots \ldots \ldots \ldots \ldots$$
$$a_{nn}x_n = b_n$$

This process is called *elimination*. Then $x_n = b_n/a_{nn}$. Further, we substitute upwards to the equations to obtain the solution. This process is called *back substitution*.

b. Complex matrix A and vector b.

An equivalent 2nx2n system of real equations is used:

$$\begin{bmatrix} Re\{\underline{A}\} & -Im\{\underline{A}\} \\ --- & ---- \\ Im\{\underline{A}\} & Re\{\underline{A}\} \end{bmatrix} \begin{bmatrix} Re\{\underline{x}\} \\ --- \\ Im\{\underline{x}\} \end{bmatrix} = \begin{bmatrix} Re\{\underline{b}\} \\ --- \\ Im\{\underline{b}\} \end{bmatrix}$$

```
                    LINEQ                      LINEQ
                   LINEQ                      LINEQ
                  LINEQ                      LINEQ
                 LINEQ                      LINEQ
                LINEQ                      LINEQ
               LINEQ                      LINEQ
              LINEQ                      LINEQ
      Solution of a system of linear equation with the Gauss Elimination Method
                       LINEQ      LINEQ
                      LINEQ    LINEQ
                     LINEQ LINEQ
                    LINELINEQ
                   LILINEQ
                   LINEQ

    MENU

    <Q>  Quit
    <L>  Load File
    <I>  Input Data
    <S>  Save file
    <C>  Compute
    <P>  Print Results

    Enter Your Selection?
```

The results vector has the real part of the solution in the first n-elements and the imaginary part in the remaining elements.

Use: Module LINEQ can be used in developed programs or as self-standing one to solve the problem with data stored in a text file.

To use it as a self-standing program, select LINEQ from COMPUME.

The first page has a command menu with 4 program commands: **Q, P, I, S**.

Q quits the program execution and returns to COMPUME.

I initiates an interactive section to input or edit data and prepare the data file LINEQ.DAT. It loads the existing data file from disk, LINEQ.DAT is the default file, supplied with the program diskettes.

The first line is a problem identification, i.e.,

" Test Problem"

The second line has the number of equations, i.e.,

3

The third line is an identifier,

Coefficient matrix:

Lines $i + 3$, $i = 1, 2, ..., n$, have the elements of the corresponding line of matrix a_{i1}, a_{i2}, ... ,a_{in}, separated by commas.

Line $n + 4$ is an identifier,

Constant vector:

Line $n + 5$ has the elements of the constant vector $b(i)$, $i = 1, 2, 3, ..., n$, separated by commas.

The user can modify the data interactively.

Finally, the program saves the data in the data file LINEQ.DAT.

S invokes the module GAUSSEL which reads the data file LINEQ.DAT, solves the system of linear equations and stores the data and the solution vector in the file LINEQ.RES.

P prints the contents of th LINEQ.RES file.

D computes the determinant of the matrix with Laplace expansion.

To use the module as a subprogram in user developed programs,

a. Write your-own program to input or to compute the coefficient matrix **A** and constant vector **b**. Store them in a text file LINEQ.DAT as above.

. b. The last line of your program should run the file GAUSSEL.EXE. You should consult for this your Language Reference Manual. If you work in Quick Basic, for example, your last line should be:

RUN "GAUSSEL"

c. Compile your program into a file LINEQ.EXE.

Add LINEQ.EXE in your **MELAB** root directory (it will destroy the previous LINEQ.EXE) or use it in another directory. You are now ready to use LINEQ. The module GAUSSEL reads the LINEQ.DAT file, computes the solution, stores the results in file LINEQ.RES and returns control to the LINEQ module.

To use the module GAUSSEL as a subprogram,

a. Prepare in your editor a text file with the coefficient matrix **A** and constant vector **b**, as above, and store it as LINEQ.DAT in the current drive.

b. With the GAUSSEL.EXE module also in the current dive, type,

C:\COMPUME>GAUSSEL

The module GAUSSEL reads the LINEQ.DAT file, computes the solution, stores the results in file LINEQ.RES and returns control to the LINEQ module. You can view or print the results file with your editor.

ATTENTION: Every time the module runs, destroys the old LINEQ.DAT and LINEQ.RES files to store the new ones. If you want to save them for later use, you need to rename them, i.e.

C:\COMPUME>rename lineq.dat lineq37.dat

Example: Let file LINEQ.DAT be:

```
Test data for LINEQ
 3
Coefficient matrix
 3 , 1 , 1
 1 , 3 , 4
 1 , 4 , 2
Constant vector
 8 , 19 , 15
```

Select LINEQ from COMPUME. Select **I** to read the LINEQ.DAT file and view its contents. On every entry, hit ENTER to keep the default values, or enter another value of your choice. The program returns to the command line.

Select **S** and when the command line reappears, select **P**. The result will be

```
Results from LINEQ
 3
Solution vector
 1 , 2 , 3
Coefficient matrix
 3 , 1 , 1
 1 , 3 , 4
 1 , 4 , 2
Constant vector
 8 , 19 , 15
```

and will be stored in this form in the file LINEQ.RES.

Probable errors: Division by zero or **overflow** would indicate a singular coefficient matrix, such as one with two equal rows. **File not found** indicates that LINEQ.DAT or GAUSSEL.EXE are not on the proper drive. **Subscript out of range** indicates that you have tried to solve more than 20 equations, which is the limit for this version.

2.2. **MAT**rix **INV**ersion

Function: Inversion of a real square matrix with the method of Gauss Elimination.

References: Faddeeva, V.N., 1959, Meirovitch, L., 1980, Dimarogonas 1988.

Hardware Requirements: 512k, 1 FD 360k, any monitor card, monitor.

Limitations: Matrix rank up-to 20.

Files needed: MATINVER.EXE, MATINV.EXE, MATINVER.DAT

Method: Assume a system of linear equations

$$\mathbf{A}x_i = b_i$$

where $\mathbf{b}_i = \{\ 0\ \ 0\ \ 0\ ....\ \ 1\ ...0\ \ 0\}$ and the 1 is the element i. The matrix $[x_1\ \ x_2\ ...\ \ x_n]$ is the inverse of the matrix $\mathbf{A}$. The vectors x_i are obtained with the method of Gauss elimination (see MATINVER).

Use: Module MATINVER can be used in developed programs or as self-standing one to solve the problem with data stored in a text file.

To use it as a self-standing program, select MATINVER from COMPUME.

The first page has a command menu with 4 program commands: **Q**, **P**, **I**, **S**.

Q   quits the program execution and returns to COMPUME.

I   initiates an interactive section to input or edit data and prepare the data file MATINVER.DAT. It loads the existing data file from disk, MATINVER.DAT is the default file, supplied with the program diskettes.

The first line is a problem identification, i.e.,

" Test Problem"

The second line has the rank of the matrix, i.e.,

3

The third line is an identifier,

Matrix:

Lines $i + 3$, $i = 1, 2, ..., n$, have the elements of the corresponding line of matrix a_{i1}, a_{i2}, ... ,a_{in}, separated by commas.

The user can modify the data interactively.

Finally, the program saves the data in the data file MATINVER.DAT.

S invokes the module MATINV which reads the data file MATINVER.DAT, solves the system of linear equations and stores the data and the solution vector in the file MATINVER.RES.

P prints the contents of the MATINVER.RES file.

To use the module as a subprogram in user developed programs,

a. Write your-own program to input or to compute the coefficient matrix $\mathbf{A}$ and constant vector $\mathbf{b}$. Store them in a text file MATINVER.DAT as above.

b. The last line of your program should run the file MATINV.EXE. You should consult for this your Language Reference Manual. If you work in Quick Basic, for example, your last line should be:

RUN "MATINV"

c. Compile your program into a file MATINVER.EXE.

Add MATINVER.EXE in your **MELAB** root directory (it will destroy the previous LINEQ.EXE) or use it in another directory. You are now ready to use MATINVER. The module MATINV reads the MATINVER file, computes the solution, stores the results in file MATINVER.RES and returns control to the MATINVER module.

To use the module MATINV as a subprogram,

a. Prepare in your editor a text file with the coefficient matrix **A** and constant vector **b**, as above, and store it as MATINVER.DAT in the current drive.

b. With the MATINV.EXE module also in the current dive, type,

C:\COMPUME>MATINV

The module MATINV reads the MATINVER.DAT file, computes the solution, stores the results in file MATINVER.RES and returns control to the MATINVER module. You can view or print the results file with your editor.

ATTENTION: Every time the module runs, destroys the old MATINVER.DAT and MATINVER.RES files to store the new ones. If you want to save them for later use, you need to rename them, i.e.

C:\COMPUME>rename MATINVER.dat MATINVER37.dat

Example: Let file MATINVER.DAT be:

```
Test data for MATINVER
 3
Coefficient matrix
 1 , 0 , 0
 0 , 2 , 0
 0 , 0 , 4
```

Select MATINVER from COMPUME. Select **I** to read the MATINVER.DAT file and view its contents. On every entry, hit ENTER to keep the default values, or enter another value of your choice. The program returns to the command line.

Select **S** and when the command menu reappears, select **P**. The result will be

```
Results from MATINVER
 3
Solution matrix
 1 , 0 , 0
 0 ,.5 , 0
 0 , 0 , .25
```

and will be stored in this form in the file MATINVER.RES.

Probable errors: Division by zero or **overflow** would indicate a singular coefficient matrix, such as one with two equal rows. **File not found** indicates that MATINVER.DAT or MATINV.EXE are not on the proper drive. **Subscript out of range** indicates that you have tried to solve more than 20 equations, which is the limit for this version.

2.3. Eigenvalue problem with STODOLA's method

Function: Solution of the general eigenvalue problem $\omega^2 \mathbf{M}\mathbf{x} = \mathbf{K}\mathbf{x}$, where $\mathbf{K}$ and $\mathbf{M}$ are real, symmetric matrices, with power iteration.

References: Faddeeva, 1959, Meirovitch, L., 1980, Dimarogonas & Haddad, 1992.

Hardware Requirements: 512k, 1 FD 360k, any monitor card, monitor.

Limitations: Up to 20 equations.

Files needed: Uses EIGSTOD.EXE, STODOLA.DAT, STODOLA.RES.

Method: Assume the eigenvalue problem in the form

$$\underline{K}\underline{x} = -\omega^2 \underline{M}\underline{x}$$

Let $\underline{x}_{(0)} = \{1\ 1\ ...\ 1\}$. Form the sequence:

$$\underline{K}\underline{x}_{(1)}^* = -\omega^2 \underline{M}\underline{x}_{(0)}$$

$$\underline{x}_{(1)} = \lambda_1 \underline{x}_{(1)}^* \quad \text{so that always } x_{1(j)} = 1$$

$$\underline{K}\underline{x}_{(2)}^* = -\omega^2 \underline{M}\underline{x}_{(1)}$$

$$\underline{x}_{(2)} = \lambda_2 \underline{x}_{(2)}^*$$

$$\cdot \quad \cdot \quad \cdot \quad \cdot \quad \cdot \quad \cdot \quad \cdot \quad \cdot \quad \cdot$$

$$\underline{K}\underline{x}_{(k)}^* = -\omega^2 \underline{M}\underline{x}_{(k-1)}$$

$$\underline{x}_{(k)} = \lambda_k \underline{x}_{(k)}^*$$

λ_k converges to ω_1^2 and $\underline{x}_{(k)}$ to $\underline{x}_1$. For higher modes, iterate on

$$\underline{x}_{(j)} = (1/\omega_j)^2 \underline{D}_{(j)} \underline{x}_{(j-1)}$$

where

$$\underline{D}_{(j)} = \underline{D}_{(j-1)} - (1/\omega_{j-1})^2 \underline{x}_{j-1}{}^T \underline{x}_{j-1} \underline{M}$$

$$\underline{D}_{(1)} = \underline{M}^{-1}\underline{K}$$

Use: Module STODOLA can be used in developed programs or as self-standing one to solve the problem with data stored in a text file.

To use it as a self-standing program, select STODOLA from COMPUME.

The first page has a command menu with 4 program commands: **Q, P, I, S**.

Q quits the program execution and returns to COMPUME.

I initiates an interactive section to input or edit data and prepare the data file STODOLA.DAT. It loads the existing data file from disk, STODOLA.DAT is the default file, supplied with the program diskettes.

The first line is a problem identification, i.e.,

" Test Problem"

The second line has the matrix size. The third line is an identifier,

Stiffness matrix:

Lines $i + 3$, $i = 1, 2, ..., n$, have the elements of the corresponding line of matrix k_{i1}, k_{i2}, ... , k_{in}, separated by commas.

Line $n + 4$ is an identifier,

Mass matrix:

Lines $i + n + 5$, $i = 1, 2, 3, ... , n$ have the elements of the mass matrix $m(i,j)$, $j = 1, 2, 3, ..., n$, separated by commas.

Line $2n+6$ has the number of desired eigenvalues _ n.

The user can modify the data interactively.

Finally, the program saves the data in the data file STODOLA.DAT.

S invokes the module EIGSTOD which reads the data file STODOLA.DAT, solves the system of linear equations and stores the data and the solution vector in the file STODOLA.RES.

P prints the contents of the STODOLA.RES file.

To use the module as a subprogram in user developed programs,

a. Write your-own program to input or to compute the matrices **K** and **M**. Store them in a text file STODOLA.DAT as above.

b. The last line of your program should run the file EIGSTOD.EXE. You should consult for this your Language Reference Manual. If you work in Quick Basic, for example, your last line should be:

RUN "EIGSTOD"

c. Compile your program into a file STODOLA.EXE.

Add STODOLA.EXE in your **MELAB** root directory (it will destroy the previous STODOLA.EXE) or use it in another directory. You are now ready to use STODOLA. The module EIGSTOD reads the STODOLA.DAT file, computes the solution, stores the results in file STODOLA.RES and returns control to the STODOLA module.

To use the module EIGSTOD as an independent program,

a. Prepare in your editor a text file with matrices **K** and **M**, as above, and store it as STODOLA.DAT in the current drive.

b. With the EIGSTOD.EXE module also in the current dive, type,

C:\COMPUME>EIGSTOD

The module EIGSTOD reads the STODOLA.DAT file, computes the solution, stores the results in file STODOLA.RES and returns control to the STODOLA module. You can view or print the results file with your editor.

ATTENTION: Every time the module runs, destroys the old STODOLA.DAT and STODOLA.RES files to store the new ones. If you want to save them for later use, you need to rename them, i.e.

C:\COMPUME>rename STODOLA.dat STODOLA37.dat

Example: Let file STODOLA.DAT be:

```
Test data for STODOLA
 3
Stiffness Matrix
 4 , 2 , 2
 2 , 5 , 1
 2 , 1 , 6
Mass Matrix
 1 , 0 , 0
 0 , 1 , 0
 0 , 0 , 1
 3
```

Select STODOLA from COMPUME. Select **I** to read the STODOLA.DAT file and view its contents. On every entry, hit ENTER to keep the default values, or enter another value of your choice. The program returns to the command line.

Select **S** and when the command line reappears, select **P**. The result will be

```
Results from STODOLA
STODOLA eigenvalue problem results: No of Frequencies, DOF
 3 , 3
Eigenvalues (om^2) of last problem
 2.127009 , 4.489367 , 8.387621 ,
Normal modes of last problem
 .8283801 , .1519333 , .5384992 ,
-.468117 , .7193629 , .514662 ,
-.3076571 ,-.6778151 , .6671894 ,
Mass Matrix of last problem
 1 , 0 , 0 ,
 0 , 1 , 0 ,
 0 , 0 , 1 ,
Stiffness  Matrix of last problem
 4 , 2 , 2 ,
 2 , 5 , 1 ,
 2 , 1 , 6 ,
```

and will be stored in this form in the file STODOLA.RES.

Possible Errors: Overflow, **division by zero**: Improper matrices, consult references for conditions.

Higher eigenvalues might be in error due to round-off errors.

2.4. Eigenvalue problem with **JACOBI**'s method

Function: Find the eigenvalues of a real, symmetric matrix with the Jacobi method of matrix iteration.

References: Faddeeva 1959, Meirovitch, L., 1980, Dimarogonas & Haddad, 1992

Hardware Requirements: 512k, 1 FD 360k, any monitor card, monitor.

Limitations: Up to 20 equations.

Files needed: Uses EIGENJAC.EXE, JACOBI.DAT, JACOBI.RES.

Method: Assume the eigenvalue problem in the form

$$-\omega^2 \underline{I}x + \underline{A}x = \underline{0}$$

where $\underline{A}$ a real symmetric matrix nxn. Iterate for i,j = 1, 2, ..., n:

$$\tan 2\theta = 2a_{12}/(a_{11}-a_{22})$$

```
Column:     1 2 3 .   i  . . .   j  . . n   Row:
```

$$
\underline{R} = \begin{bmatrix}
1 & 0 & 0 & . & . & & . & . & . & & . & . & 0 & 0 \\
0 & 1 & 0 & . & . & & . & . & . & & . & . & 0 & 0 \\
. & . & . & . & & . & & . & . & . & & . & . & . \\
. & . & . & . & \cos\theta & & . & . & . & \sin\theta & & . & . & . \\
. & . & . & . & . & . & & . & . & . & . & & . & . \\
0 & 0 & . & . & . & & 0 & 1 & 0 & & . & . & 0 & 0 \\
. & . & . & . & . & & . & . & . & & . & . & . \\
. & . & . & . & -\sin\theta & & . & . & . & \cos\theta & & . & . & . \\
. & . & . & . & . & . & & . & . & . & . & & . & . \\
. & . & . & . & . & & . & . & . & & . & . & . \\
0 & 0 & 0 & . & . & & . & . & . & & . & . & 1 & 0 \\
0 & 0 & 0 & . & . & & . & . & . & & . & . & 0 & 1
\end{bmatrix}
\begin{matrix}
1 \\ 2 \\ . \\ i \\ . \\ . \\ . \\ j \\ . \\ . \\ n-1 \\ n
\end{matrix}
$$

$$\underline{A}^* = \underline{R}^T \underline{A} \underline{R}$$
$$\underline{R}^* = \underline{R}^* \underline{R}$$

until the off diagonal elements of $\underline{A}^*$ become less than a prescribed number. Then, the eigenvalues ω^2 are the diagonal elements of $\underline{A}^*$ and the eigenvectors the columns of the matrix $\underline{R}^*$.

Use: Module JACOBI can be used in developed programs or as self-standing one to find the eigenvalues of a matrix stored in a text file.

To use it as a self-standing program, select JACOBI from COMPUME.

The first page has a command menu with 4 program commands: **Q, P, I, S**.

Q quits the program execution and returns to COMPUME.

I initiates an interactive section to input or edit data and prepare the data file JACOBI.DAT. It loads the existing data file from disk, JACOBI.DAT is the default file, supplied with the program diskettes.

The first line is a problem identification, i.e.,

" Test Problem"

The second line has the number of equations, i.e.,

3

The third line is an identifier,

matrix:

Lines i + 3, i = 1, 2, ..., n, have the elements of the corresponding line of matrix a_{i1}, a_{i2}, ... ,a_{in}, separated by commas.

The user can modify the data interactively.

Finally, the program saves the data in the data file JACOBI.DAT.

S invokes the module EIGENJAC which reads the data file JACOBI.DAT, solves the system of linear equations and stores the data and the solution vector in the file JACOBI.RES.

P prints the contents of th JACOBI.RES file.

To use the module as a subprogram in user developed programs,

a. Write your-own program to input or to compute the coefficient matrix **A** and constant vector **b**. Store them in a text file JACOBI.DAT as above.

b. The last line of your program should run the file EIGENJAC.EXE. You should consult for this your Language Reference Manual. If you work in Quick Basic, for example, your last line should be:

RUN "EIGENJAC"

c. Compile your program into a file JACOBI.EXE.

Add JACOBI.EXE in your **MELAB** root directory (it will destroy the previous JACOBI.EXE) or use it in another directory. You are now ready to use JACOBI. The module EIGENJAC reads the JACOBI.DAT file, computes the solution, stores the results in file JACOBI.RES and returns control to the JACOBI module.

To use the module EIGENJAC as a subprogram,

a. Prepare in your editor a text file with the coefficient matrix **A** and constant vector **b**, as above, and store it as JACOBI.DAT in the current drive.

b. With the EIGENJAC.EXE module also in the current dive, type,

C:\COMPUME>EIGENJAC

The module EIGENJAC reads the JACOBI.DAT file, computes the solution, stores the results in file JACOBI.RES and returns control to the JACOBI module. You can view or print the results file with your editor.

ATTENTION: Every time the module runs, destroys the old JACOBI.DAT and JACOBI.RES files to store the new ones. If you want to save them for later use, you need to rename them, i.e.

C:\COMPUME>rename JACOBI.dat JACOBI37.dat

Example: Let file JACOBI.DAT be:

```
Test data for Jacobi iteration, matrix dimension
 3
Matrix
 4 , 2 , 2
 2 , 5 , 1
 2 , 1 , 6
 0 , 0 , 0
```

Select JACOBI from COMPUME. Select **I** to read the JACOBI.DAT file and view its contents. On every entry, hit ENTER to keep the default values, or enter another value of your choice. The program returns to the command line.

Select **S** and when the command line reappears, select **P**. The result will be

```
Results of Jacobi Eigenvalue Solution
 3
```

```
----------Eigenvalues-----------
 2.125924 , 4.486456 , 8.387618
-------------eigenvectors----------------
-.8280333 , .1555202 , .5386782
 .4696546 , .7171605 , .5148838
 .3062439 ,-.6793336 , .6668736
Matrix of last problem
 4 , 2 , 2
 2 , 5 , 1
 2 , 1 , 6
```

and will be stored in this form in the file JACOBI.RES.

Possible Errors: Overflow: Improper matrices, consult references for conditions.

Higher modes might be in error due to round-off errors.

2.5. Eigenvalue problem with the **CHOLESKY**'s method

Function: Solution of the eigenvalue problem with the Cholesky method of matrix decomposition and Jacobi iteration.

References: Bendot & Cholesky 1924, Banachiewicz 1938, Dimarogonas & Haddad, 1992.

Hardware Requirements: 512k, 1 FD 360k, any monitor card, monitor.

Limitations: Up to 20 equations.

Files needed: Uses EIGCSKY.EXE, CHOLESKY.DAT, CHOLESKY.RES.

Method: Assume the eigenvalue problem in the form

$$(-\omega^2 \underline{M} + \underline{K})\underline{x} = 0$$

Set

$$\underline{D} = \underline{M}^{-1}\underline{K} = \underline{U}^T\underline{U}$$

where

$$u_{11} = (d_{11})^{1/2}$$
$$u_{1j} = (d_{1j})/u_{11}, \quad j=2,3,\ldots,n$$

$$u_{ii} = (d_{ii} - \sum_{j=1}^{i-1} u^2_{ji})^{1/2}, \quad i = 2, 3, \ldots, n$$

$$u_{ij} = (d_{ij} - \sum_{k=1}^{i-1} u_{ki}u_{kj}), \quad \begin{array}{l} i = 2, 3, \ldots, n, \\ j = i+1, i+2, \ldots, n \end{array}$$

$$u_{ij}=0, \quad i>j$$

$$\underline{K}=\underline{U}^T\underline{U}$$
$$\underline{B}=(\underline{U}^T)^{-1}\underline{M}\underline{U}^{-1}$$

$$\omega^2 \underline{B}\underline{y} = \underline{y}$$

Find eigenvalues $(\omega_i)^2$ and eigenvectors $\underline{y}_i$ of the above auxiliary problem with Jacobi method. Then

$$\underline{u}^{-1} = [y_{ij}],$$
$$y_{ii} = 1/u_{ii}$$

$$y_{ij} = -\left(\sum_{k=i+1}^{j} u_{ik}y_{kj} \right)/u_{ii}, \quad i < j$$

$$y_{ij} = 0, \quad i > j$$

The eigenvectors are:

$$\underline{x} = \underline{u}^{-1}\underline{y}$$

Use: Module CHOLESKY can be used in developed programs or as self-standing one to solve the problem with data stored in a text file.

To use it as a self-standing program, select CHOLESKY from COMPUME.

The first page has a command menu with 4 program commands: **Q, P, I, S**.

Q quits the program execution and returns to COMPUME.

I initiates an interactive section to input or edit data and prepare the data file CHOLESKY.DAT. It loads the existing data file from disk, CHOLESKY.DAT is the default file, supplied with the program diskettes.

The first line is a problem identification, i.e.,

" Test Problem"

The second line has the matrix size.

The third line is an identifier,

matrix K:

Lines $i + 3$, $i = 1, 2, ..., n$, have the elements of the corresponding line of matrix $k_{i1}, k_{i2}, ..., k_{in}$, separated by commas.

Line $n + 4$ is an identifier,

matrix M:

Lines $n + 4 + i$, $i = 1, 2, 3, ..., n$ have the elements of the mass matrix $m(i,j)$, $j = 1, 2, 3, ..., n$, separated by commas.

The user can modify the data interactively.

Finally, the program saves the data in the data file CHOLESKY.DAT.

S invokes the module EIGCSKY which reads the data file CHOLESKY.DAT, solves the system of linear equations and stores the data and the solution vector in the file CHOLESKY.RES.

P prints the contents of th CHOLESKY.RES file.

To use the module as a subprogram in user developed programs,

a. Write your-own program to input or to compute the matrix **A**. Store it in a text file CHOLESKY.DAT as above.

b. The last line of your program should run the file EIGCSKY.EXE. You should consult for this your Language Reference Manual. If you work in Quick Basic, for example, your last line should be:

RUN "EIGCSKY"

c. Compile your program into a file CHOLESKY.EXE.

Add CHOLESKY.EXE in your **MELAB** root directory (it will destroy the previous CHOLESKY.EXE) or use it in another directory. You are now ready to use CHOLESKY. The module EIGCSKY reads the CHOLESKY.DAT file, computes the solution, stores the results in file CHOLESKY.RES and returns control to the CHOLESKY module.

To use the module EIGCSKY as a subprogram,

a. Prepare in your editor a text file with the matrix **A** , as above, and store it as CHOLESKY.DAT in the current drive.

b. With the EIGCSKY.EXE module also in the current dive, type,

C:\COMPUME>EIGCSKY

The module EIGCSKY reads the CHOLESKY.DAT file, computes the solution, stores the results in file CHOLESKY.RES and returns control to the CHOLESKY module. You can view or print the results file with your editor.

ATTENTION: Every time the module runs, destroys the old CHOLESKY.DAT and CHOLESKY.RES files to store the new ones. If you want to save them for later use, you need to rename them, i.e.

C:\COMPUME>rename CHOLESKY.dat CHOL37.dat

Example: Let file CHOLESKY.DAT be:

```
Test data for Cholesky
 3
Stiffness Matrix
 4 , 2 , 2
 2 , 5 , 1
 2 , 1 , 6
Mass Matrix
 1 , 0 , 0
 0 , 1 , 0
 0 , 0 , 1
```

Select CHOLESKY from COMPUME. Select **I** to read the CHOLESKY.DAT file and view its contents. On every entry, hit ENTER to keep the default values, or enter another value of your choice. The program returns to the command line.

Select **S** and when the command line reappears, select **P**. The result will be

```
Cholesky method: last problem results
 3
Eigenvalues
 2.125924 , 4.486456 , 8.387619
Eigenvectors
-.8280334 , .1555202 , .5386782
 .4696546 , .7171606 , .5148838
```

```
 .306244 , -.6793337 , .6668736
Stiffness  Matrix of last problem
 4 , 2 , 2
 2 , 5 , 1
 2 , 1 , 6
Mass Matrix of last problem
 1 , 0 , 0
 0 , 1 , 0
 0 , 0 , 1
```

and will be stored in this form in the file CHOLESKY.RES.

Possible Errors: Division by zero, Overflow: Improper matrices, consult references for conditions.

Higher modes might be in error due to round-off errors.

2.6 Numerical Integration of **NONLIN**ear differential equations

Function: Numerical integration of differential equations with different methods.

Available Code: Microsoft .OBJ, .EXE

References: Meirovitch, L., 1980, Dimarogonas & Haddad, 1992.

Hardware Requirements: 512k, 1 FD 360k, any monitor card, monitor.

Limitations: Up to 20 equations.

Method: System of nonlinear differential equations:

$$\underline{x}^{\circ} = \underline{f}(\underline{x}, t)$$
$$\underline{x} = \{x_1\ x_2\ \cdots\ x_n\},\quad \underline{f} = \{f_1\ f_2\ \cdots\ f_n\}$$

<u>Euler Method:</u>
Predictor: $\underline{x}_1(t+h) = \underline{x}_0 + \underline{f}(\underline{x}, t)h$

Corrector: $\underline{x}_2(t+h) = \underline{x}_0 + h[\underline{f}(\underline{x}_0, t) + \underline{f}(\underline{x}_1, t+h)]/2$

<u>Runge-Kutta Method:</u>
$$\underline{x}(t+h) = \underline{x}(t) + (\underline{K}_1 + 2\underline{K}_2 + 2\underline{K}_3 + \underline{K}_4)/6$$

where

$$\underline{K}_1 = h\underline{f}(\underline{x}_0, t)$$
$$\underline{K}_2 = h\underline{f}(\underline{x}_0 + \underline{K}_1/2, t+h/2)$$
$$\underline{K}_3 = h\underline{f}(\underline{x}_0 + \underline{K}_1/2 + \underline{K}_2/2, t+h/2)$$
$$\underline{K}_4 = h\underline{f}(\underline{x}_0 + \underline{K}_3, t+h)$$

<u>Adams-Moulton Method:</u>
Predictor: $\underline{u}_{n+1} = \underline{x}_n + (h/24)[-9\underline{f}(\underline{x}_{n-3}, t_{n-3}) + 37\underline{f}(\underline{x}_{n-2}, t_{n-2}) -$
$- 59\underline{f}(\underline{x}_{n-1}, t_{n-1}) + 55\underline{f}(\underline{x}_n, t_n)]$

Corrector: $\underline{x}_{n+1} = \underline{x}_n + (h/24)[\underline{f}(\underline{x}_{n-2}, t_{n-2}) - 5\underline{f}(\underline{x}_{n-1}, t_{n-1}) +$
$+ 19\underline{f}(\underline{x}_n, t_n) + 9\underline{f}(\underline{u}_{n+1}, t_{n+1})]$

Use: Subroutine NONLIN can be used in developed programs and it is not self-standing. It is part of the program POINCARE and it will be further discussed later.

3

CHAPTER THREE
COMPUTER GRAPHICS

3.1. SOLID

Function: Creates, synthesizes and plots solids.

Reference: Dimarogonas 1988, chapter 3.

Hardware Requirements: 512k, 1 FD 360k, EGA card, color or monochrome monitor.

Files needed: SOLID.EXE, COMPUME.SYS, data files *.SOL

Limitations: Up to 2500 nodes, 2500 lines.

Method: The solid is described by way of plane surface polygons. Circular arcs are described as inscribed polygons. Solids are created,

a. Patch-by-patch by separate definition of each surface polygon.

b. By sweeping the space by plane polygons.

c. By Boolean operations on separate solids and constructive solid geometry.

d. By the solid operations, scaling, translation, rotation.

Use: To use the program (User types what below is underlined):

1. Select from COMPUME or type <u>SOLID</u>.

2. The first window is for program identification and basic data, maximum y or height of screen in the units to be used in the solid dimensions, or the maximum vertical distance expected in the solid. Enter ymax and hit the ENTER key.

3. Next is then the main menu window.

4. You are now in the main menu. You select with the keypad arrows. The > sign indicates the selection and at the bottom line there is an explanation of the command. ENTER invokes the respective command. Alternatively, just type the first letter of the command.

Q quits the program execution.

R rotates the solid about a line through its geometric center having direction cosines Ix, Iy in respect to the x and y axes respectively and by an angle f, in radians. Enter Ix,Iy,f and ENTER at the ? prompt. Attention: Ix^2+Iy^2 should be <=1. A good axonometric view is obtained by entering .7,.7,.7

T translates the solid by Tx,Ty,Tz in respect to its present position on the xyz system.

E scales the solid by multiplying all coordinates x,y,z by Sx,Sy,Sz respectively. Enter these values at the ? prompt.

A scales the solid so that it occupies the rectangle 0<x<xmax and 0<y<ymax. Enter xmax, ymax at the ? prompt.

P plots the solid currently in memory on the screen. There are three choices: <1> solid view with all hidden lines removed, <2> solid view with a quick hidden line algorithm which works right only for convex solids, <3> wireframe view of all solid lines. Make selection and hit ENTER. Selection 1 might take long time for complex solids with curved surfaces.

N starts a new drawing. Any previous solid is erased from the operation register.

S saves the current solid on disc. At prompt, enter file name desired (preferable with a .sol extension). If you use wild characters, it lists the files specified on the current drive. I.e. *.sol prints a list of files with *.sol extension for reference. If no name is entered, you return to menu.

L loads a solid from disk. If you have a solid already in the operation register, you add to it the new solid. See above for name conventions.

C clears the screen without erasing the solid from memory.

F flops screens. There are 3 screens on which one can plot different drawings. Enter the screen desired. Availability of the screens depends on your EGA card memory.

B manipulates memory registers on which one can store different solids or add solids to form one combined solid. If solids are added, the program finds also the surface intersection lines.

D deletes last polygon created with Make Solid command.

M makes a new solid: It invokes the model menu:

N defines a node. This can be a start of a new polygon or not. Enter appropriate selection.

L plots a line from the previous defined node or from the last point plotted, whichever is more recent. It can be a full line or a fantom line. This is the side of a polygon we do not want to be visible. For example, if a polygon has a hole, we define a continuous polygon with fantom lines connecting the inner and outer polygons. The inner polygon is traced in the inverse direction.

G plots a grid on the screen for user reference.

A plots an arc as an open polygon. The arc can be defined in two ways: a) Define three nodes and select the three point selection after you hit A. An arc is then drawn from the first to the third going through the second point. b) Define two nodes and hit A. Then select the two point option. The program asks for the start and end angle. After you enter them, it plots an arc with center the first node and radius the distance of the two nodes starting from the start angle up to the end angle. Attention: The sign of the angles (+ CW) should be carefully observed, if the direction of the polygon tracing is important (see below).

The following two commands are related to the motion of the cross cursor which moves with the keypad arrows.

S move the cursor in smaller steps (division by 10)

F move the cursor in larger steps (multiplication by 10)

P define a polygon after you return to its starting node. If the convex solid plot option will be used, the sequence of nodes, lines and arc definitions should be made in the positive direction as we look from outside the solid we will create (Clockwise).

CAUTION: Always close a polygon with the P command. If you do not, the program assumes that the polygon continues and is mixed up with the next one.

D displace a polygon. This command invokes the displace menu, which includes: Scaling, rotation, translation. These operations are performed about the cross cursor. Thus, a polygon can be scaled and rotated before is translated. Before you use them, place the cross in the desired position. Translation is performed with the keypad direction keys. The amount of translation equals the step of the movement of the cursor.

This can be changed with the S and F commands. When you finish translation, return to Make Model Menu.

C Connect the last two polygons node by node creating a solid. The second polygons was created with the displace command or independently. In the latter case, the two polygons must have the same number of nodes.

E Exit to the main menu.

X display the front view, yz plane.

Y display the right view, xz plane.

Z display the top view, xy plane.

R Rotate in the current view **about cursor** + . Enter angle.

T Translate in the current view. Use direction keys.

M Magnify (Scale) on current view. Enter scale factor.

P Push a specific node to a new location.

CREATING A SOLID:

Prismatic solids can be created by defining the base polygon and then displacing, scaling or rotating it. To create hollow sections first create the outer polygon. Close but do not use P command and move with a fantom line to the first node of the inner polygon. Create the inner polygon in the opposite direction. When you close it, do not use the P command also. Move to the first node of the outer polygon with a fantom line. Then, hit P and define the polygon which includes both the inner and outer one. This polygon can now be displaced and connected to create the solid.

Individual solids can be stored on disc after they have the proper position in the three views or they can be stored on different registers from the main menu.

To assembly a complex system or solid, prepare each element and place it in the right position, place them on different files on disk. Load them one after the other and put them on different registers. Use the B command to manipulate registers and add solids. For example, to connect solids A and B:

a. Create solid A using Make Model. Type M to return to the main MENU. Then type B. The Register manipulation page appear on the screen. The solid A is on register 0 (the operation register). Add it to the register 1. Erase register 0. Now solid A is in storage register 1 and the operation register 0 is clear. Return to the main menu. Select Make Model Menu. You see solid A in different color while you create solid B. Upon solid operations, solid A is not affected because it is not in the operations register 0. Manipulate solid B until it is in the desired relation to A in all 3 screens. Exit to the main menu. Type B to go to the register manipulation menu. Add register 1 (solid A) to the register 0 (solid B). The program asks if you want the intersection of the two solids. If the answer is Y(es) and the solids intersect, the intersection lines are drawn. Clear register 1. Exit to the main menu. Save and plot the combined solid.

Example I: Draw a double T section with height 120 mm, width 60 mm, web and flange thickness 8 mm.

SOLID is selected from the COMPUME. Select Make Solid from the main menu. Hit Z. You are now in the (x-y) plane. Move cursor to (0,0) and hit N. Answer YES to the prompt asking if you start a new polygon. Moving counter-clockwise, move the cursor successively to the corners defining the section hitting L each time to draw the respective edge from the last point. The position of the cursor is the origin (0,0). Hit L and then P, to define a polygon. Hit M to return to the main menu. Hit S to save with a name of your choice.

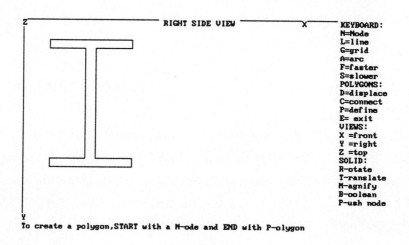

Example II: A cube with an inner rectangular opening is modelled with SOLID using <Make Solid>. The solid is shown below as a wireframe and a solid model with the hidden lines removed:

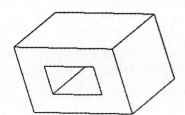

```
                                                    MAIN MENU

                                                    Quit
                                                    Rotate
                                                    Translate
                                                    Expand
                                                    Autoscale
                                                   >Plotscreen
                                                    New
                                                    Save File
                                                    Load File
                                                    Clearscrn
                                                    Make Solid
                                                    Flop scr
                                                    Buffer
                                                    Delete last

                                                    COMMAND
```

```
        Plotting line  18  of polygon  13
```

The data file DCUBE.SOL follows:

```
NODE, 28.61635 , 73.71159 , 38.47727
NODE, 99.02192 , 75.64128 , 76.41582
NODE, 113.4653 , 137.2462 , 46.47871
NODE, 43.0597 , 135.3165 , 8.540167
NODE, 50.34441 , 91.7954 , 39.40845
NODE, 56.53442 , 118.1975 , 26.57826
NODE, 91.7372 , 119.1623 , 45.54753
NODE, 85.5472 , 92.76025 , 58.37772
NODE,-1.320751 , 106.9078 , 92.34545
NODE, 69.08481 , 108.8375 , 130.284
NODE, 83.52815 , 170.4424 , 100.3469
NODE, 13.12259 , 168.5127 , 62.40835
NODE,-1.320751 , 106.9078 , 92.34545
NODE, 20.40731 , 124.9916 , 93.27663
NODE, 26.59731 , 151.3937 , 80.44644
NODE, 61.80009 , 152.3586 , 99.41571
NODE, 55.61009 , 125.9565 , 112.2459
NODE, 20.40731 , 124.9916 , 93.27663
NODE,-1.320751 , 106.9078 , 92.34545
POLYGON, 11 , 1 , 2 , 3 , 4 , 1 ,-5 , 6 , 7 , 8 , 5 ,-1
POLYGON, 5 , 9 ,-10 , 2 ,-1 ,-9
POLYGON, 5 , 10 ,-11 , 3 ,-2 ,-10
POLYGON, 5 , 11 ,-12 , 4 ,-3 ,-11
POLYGON, 5 , 12 ,-13 , 1 ,-4 ,-12
POLYGON, 5 , 13 ,-14 , 5 ,-1 ,-13
POLYGON, 5 , 14 ,-15 , 6 ,-5 ,-14
```

```
POLYGON, 5 , 15 ,-16 , 7 ,-6 ,-15
POLYGON, 5 , 16 ,-17 , 8 ,-7 ,-16
POLYGON, 5 , 17 ,-18 , 5 ,-8 ,-17
POLYGON, 5 , 18 ,-19 , 1 ,-5 ,-18
POLYGON, 11 , 19 ,-18 , 17 , 16 , 15 , 14 ,-13 , 12 , 11 , 10 , 9
SOLEND
```

Probable Errors: Execution breaks: Very large solid for the screen. Hardware incompatibility, wrong graphics adaptor. Out of memory. When loading file, no such file in current drive. Lines mixed-up: You did not close a polygon with P.

3.2. AUTOmatic MESH generation

Function: Automatic generation of finite element triangular and quadrilateral mesh for Finite Element Analysis. Creates and manipulates plane Finite Element Meshes.

Reference: Dimarogonas 1988, chapter 3.

Hardware Requirements: 512k, 1 FD 360k, EGA card, monochrome monitor.

Limitations: Up to 2500 nodes, 2500 elements.

Files needed: AUTOMESH.EXE, COMPUME.SYS, MESH*.DAT data files.

Method: A curved quadrilateral is defined by a number of $n\xi \times n\eta$ points along general curves. Polynomial transformation is used to transform from the (x,y) plane to the (η,ξ) plane where the quadrilateral is transformed into a rectangle:

$$x = \sum_{i=1}^{n} [\sum_{j=1}^{m} a_{ij}\xi^{j}]\eta^{i}$$

$$y = \sum_{i=1}^{n} [\sum_{j=1}^{m} b_{ij}\xi^{j}]\eta^{i}$$

a_{ij} and b_{ij} are determined by collocation. A number of mxn points (x,y) must be specified along m numbered ξ-curves and n numbered η-curves.

The rectangle in the (η,ξ) plane is the divided into any number of elements $n_x \times n_y$. Then the inverse transformation above creates the curved elements on the (x,y) plane. Addition or subtraction and scaling can be performed with meshes to create a complex mesh.

Use: To use the program (User types what below is underlined):

1. Select from COMPUME or type AUTOMESH in your directory.

2. The first window is for program identification and basic data:

a. Type of element: triangular, quadrilateral. Enter choice

b. Maximum width x of screen in the units to be used in the mesh dimensions, or the maximum horizontal distances expected in the mesh. Enter xmax and hit ENTER key.

3. Next is then the main menu window.

4. You are now in the main menu. You select with the keypad arrows. The > sign indicates the selection and at the bottom line there is an explanation of the command. ENTER invokes the respective command.

Q quits the program execution.

N starts a new mesh. Any previous mesh is erased.

S saves the current mesh on disc. At prompt, enter file name desired (preferably MESH*.DAT). It prints a list of files MESH*.DAT for reference. If no name is entered, you return to menu.

L loads a mesh file from disk. See above.

C clears the screen without erasing the mesh from memory.

A generates the mesh. The mesh generation window appears.

N defines a node at the cross cursor.

E creates an element. Previously, you define the appropriate number of nodes (2 or 3 or 4). The E command then creates the element.

The following two commands are related to the motion of the cross cursor which moves with the keypad arrows.

F moves the cursor in larger steps (x 10)

S moves the cursor in smaller steps (division by 10)

W writes text starting from the cursor location

G plots a grid on screen for reference

A performs the automatic mesh generation

R removes redundant nodes

M Exit to the main menu.

P Push a specific node to a new location.

CREATING A MESH:

To create a curved quadrilateral mesh, define first a grid of ix by iy nodes in a row-wise sequence following the boundary of the curved quadrilateral. Then hit A. Enter ix,iy. Then the program asks for Nx,Ny which is the division of the curved quadrilateral in Nx rows of nodes, Ny nodes each. The mesh is then created. To add elements, use the E command above. To modify a node location, use the P(ush node) command.

You can create the mesh in many adjacent patches. To remove the overlapping (redundant) nodes, hit R. This option connects the nodes which are closer to one another of the tolerance specified. To connect two regions, the adjacent sides should have the same number of nodes.

Example I: A curved quadrilateral is defined by its 3 edges at right angles, of lengths 60, 85, 120 mm, the first two defining a Cartesian coordinate system (x,y). The fourth edge is curved and it is defined by a point (75,30) mm and the two end points (60,0) and (120,85). Divide it in a triangular mess by a set of 8x8 nodes.

AUTOMESH is selected from COMPUME. Select, triangular from the first page, then Automesh from the main menu, then move cursor to point (0,0) and hit N to define a node. Define further nodes (37,0), (60,0), first row, (0,30), (43,30), (75,30), second row, (0,85), (62,85), (120,85), third row. Hit A, enter 3,3 and then enter 6,8. The mesh is created and plotted. Hit M to return to the main menu. Hit S to save the mesh. You are prompted to enter the file number. If you enter the wild card character * the available file names are printed with the convention used in the DIR command of DOS.

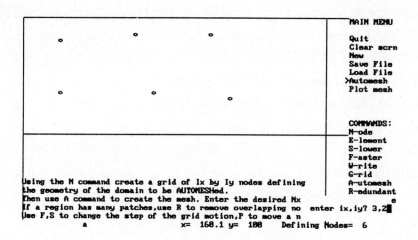

Example II: A curved quadrilateral is divided in triangular elements by way of ny = 4 horizontal and nx = 9 vertical layers of 4x9 = 36 nodes. The mesh perimeter is defined by a coarse mesh, a sequence of iy = 2 horizontal and ix = 3 vertical layers of 2x3 = 6 nodes, 'vertical' meaning from top to bottom and 'horizontal' from left to right, not necessarily along straight lines. The <Automesh> command is selected and the nodes are defined by the N command in the sequence shown below:

Then the A command creates the mesh. The user enters first ix, iy and the nx, ny. The resulting mesh and the file MESH2.DAT follow:

1 o   2 o   3 o

        4 o   ˙5 o

                                6 o
```
TRIANGULAR
NODE, 1 , 4.000002E-02 , .24
NODE, 2 , 4.020834E-02 , .4
NODE, 3 , 4.041672E-02 , .5600001
NODE, 4 , 4.062498E-02 , .7199999
NODE, 5 , .2 , .24
NODE, 6 , .2002083 , .3766667
NODE, 7 , .2004167 , .5133332
NODE, 8 , .2006249 , .6499999
NODE, 9 , .36 , .24
NODE, 10 , .3602083 , .36
NODE, 11 , .3604167 , .48
NODE, 12 , .360625 , .5999999
NODE, 13 , .52 , .2400001
NODE, 14 , .5202083 , .3500001
NODE, 15 , .5204167 , .46
NODE, 16 , .5206251 , .5700001
NODE, 17 , .68 , .2400001
NODE, 18 , .6802083 , .3466668
NODE, 19 , .6804166 , .4533334
NODE, 20 , .680625 , .5599999
NODE, 21 , .8399999 , .2400001
NODE, 22 , .8402082 , .3499999
NODE, 23 , .8404166 , .4599999
NODE, 24 , .840625 , .5699998
NODE, 25 , .9999998 , .2400001
NODE, 26 , 1.000208 , .3600001
NODE, 27 , 1.000417 , .4799998
NODE, 28 , 1.000625 , .5999999
NODE, 29 , 1.16 , .2400001
NODE, 30 , 1.160208 , .3766668
NODE, 31 , 1.160417 , .5133333
NODE, 32 , 1.160625 , .6499999
NODE, 33 , 1.32 , .2400001
NODE, 34 , 1.320208 , .4000002
NODE, 35 , 1.320416 , .5600002
NODE, 36 , 1.320625 , .72
ELEMENT, 1 , 1 , 5 , 2
ELEMENT, 2 , 5 , 6 , 2
ELEMENT, 3 , 2 , 6 , 3
ELEMENT, 4 , 6 , 7 , 3
ELEMENT, 5 , 3 , 7 , 4
ELEMENT, 6 , 7 , 8 , 4
ELEMENT, 7 , 5 , 9 , 6
ELEMENT, 8 , 9 , 10 , 6
ELEMENT, 9 , 6 , 10 , 7
ELEMENT, 10 , 10 , 11 , 7
ELEMENT, 11 , 7 , 11 , 8
ELEMENT, 12 , 11 , 12 , 8
ELEMENT, 13 , 9 , 13 , 10
```

```
ELEMENT, 14 , 13 , 14 , 10
ELEMENT, 15 , 10 , 14 , 11
ELEMENT, 16 , 14 , 15 , 11
ELEMENT, 17 , 11 , 15 , 12
ELEMENT, 18 , 15 , 16 , 12
ELEMENT, 19 , 13 , 17 , 14
ELEMENT, 20 , 17 , 18 , 14
ELEMENT, 21 , 14 , 18 , 15
ELEMENT, 22 , 18 , 19 , 15
ELEMENT, 23 , 15 , 19 , 16
ELEMENT, 24 , 19 , 20 , 16
ELEMENT, 25 , 17 , 21 , 18
ELEMENT, 26 , 21 , 22 , 18
ELEMENT, 27 , 18 , 22 , 19
ELEMENT, 28 , 22 , 23 , 19
ELEMENT, 29 , 19 , 23 , 20
ELEMENT, 30 , 23 , 24 , 20
ELEMENT, 31 , 21 , 25 , 22
ELEMENT, 32 , 25 , 26 , 22
ELEMENT, 33 , 22 , 26 , 23
ELEMENT, 34 , 26 , 27 , 23
ELEMENT, 35 , 23 , 27 , 24
ELEMENT, 36 , 27 , 28 , 24
ELEMENT, 37 , 25 , 29 , 26
ELEMENT, 38 , 29 , 30 , 26
ELEMENT, 39 , 26 , 30 , 27
ELEMENT, 40 , 30 , 31 , 27
ELEMENT, 41 , 27 , 31 , 28
ELEMENT, 42 , 31 , 32 , 28
ELEMENT, 43 , 29 , 33 , 30
ELEMENT, 44 , 33 , 34 , 30
ELEMENT, 45 , 30 , 34 , 31
ELEMENT, 46 , 34 , 35 , 31
ELEMENT, 47 , 31 , 35 , 32
ELEMENT, 48 , 35 , 36 , 32
 FILEND
```

Probable Errors: Execution breaks: Very large mesh for the screen. Hardware incompatibility, wrong graphics adaptor. Out of memory.

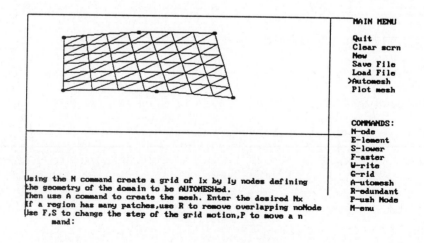

36

4

CHAPTER FOUR
STRUCTURAL ANALYSIS

4.1. BEAM STATic analysis

Function: Static Stress Analysis of Beams, Transfer Matrix Method.

Reference: Dimarogonas 1988, chapter 4.

Hardware Requirements: 512k, 1 FD 360k, EGA card, monochrome monitor.

Limitations: Up to 10 nodes, 9 elements, 3 solid supports. Limits can be extended by appropriate changes in program.

Files needed: BEAMSTAT.EXE, COMPUME.SYS, data files BEAM*.DAT.

Method: The transfer matrix method is used (see also ROTORDYN) with

$$
\begin{array}{cc}
\text{Field matrix} & \text{Node matrix} \\
[L] = \begin{bmatrix} 1 & L & L^2/2EI & L^3/2EI & 0 \\ 0 & 1 & L & L^2/2EI & 0 \\ 0 & 0 & 1 & L & 0 \\ 0 & 0 & 0 & 0 & 1 \end{bmatrix}, & [P] = \begin{bmatrix} 1 & 0 & 0 & 0 & 0 \\ 0 & 1 & 0 & 0 & 0 \\ 0 & -k_T & 1 & 0 & M_0 \\ -k & 0 & 0 & 1 & V_0 \end{bmatrix}
\end{array}
$$

State vector: $\underline{v} = \{x \ \theta \ M \ V \ 1\}$. Solution:

$$\underline{A} = \underline{P}_n \underline{L}_{n-1} \underline{P}_{n-1} \underline{L}_{n-2} \underline{P}_{n-2} \cdots \underline{P}_2 \underline{L}_1 \underline{P}_1$$

$$y_1 = (-a_{35}a_{42}+a_{32}a_{45})/(a_{31}a_{42}-a_{32}a_{41})$$

$$\theta_1 = (-a_{31}a_{45}+a_{41}a_{35})/(a_{31}a_{42}-a_{32}a_{41})$$

$$\underline{v}_j = \underline{P}_j \underline{L}_{j-1} \underline{P}_{j-1} \underline{L}_{j-2} \underline{P}_{j-2} \cdots \underline{P}_2 \underline{L}_1 \underline{P}_1 \underline{v}_1$$

Rigid supports are modeled as springs with stiffness $k = 10^3(48EI/L^3)$, where EI is the maximum flexural rigidity between the support and the closest one and L the respective length.

Use: To use the program (User types what below is underlined):

To use the compiled program, select from COMPUME or type <u>TMSTAT</u>.

Type of Section: Circular or general. If circular, the diameters have to be specified only, later on. If general, the section moment of inertia has to be specified.

File name with beam data. If none exists, hit ENTER.

Output file to save beam geometry and loading for later use.

Next is then the menu window.

One can load a file or select Make Model, which opens the data window.

Now, the data window is displayed:

Problem Identification: Any alphanumeric text.

Number of elements: Up-to 10.

Modulus of elasticity: Use a consistent system of units. SI is suggested.

Every subsequent line corresponds to a node or element numbered consecutively from left to right. Every time you hit ENTER, the cursor goes to the next entry. To enter a new value, type it and hit ENTER. For zero or to maintain the default or the previously defined value, just hit ENTER. The first 3 entries are for the element:

Length of the element, distance of node from previous one.

Diameter, for circular section, area moment of inertia for a general section.

Distributed load per unit length over the element, including the beam weight.

The next 2 entries in the line are for node data:

Lateral Force on the node

Spring constant of flexible support, if such support exists. Do not include here solid supports.

If you enter zero length or section size, the line is repeated.

The last line has node data only because there is one node more than elements.

Then the program asks for the number of solid supports. Enter their number and node numbers of their location, as the program requests them.

Finally, the program asks if data are correct. If the answer is no, the program returns in the beginning. The data entered already are printed on their position. Type only the corrections or additions. For all correct entries, just hit ENTER. You can repeat this as many times as you like.

If the data are correct, the program returns to the menu. You should save the data into a file, if you wish. Then, you proceed with the analysis. The program prints the deflection, slope, moment, shear and deflection at each node. Note the finite deflections at the supports. This is due to the way supports are modelled, by the program, as very hard springs. They have to be 2-3 orders of magnitude smaller than the other deflections. Otherwise, there is a numerical accuracy problem.

Finally, the program plots the beam, loading, deflection, bending moment, shear and their maximum values. If you respond Y to the question "CONTINUE ?", the program returns to the main menu. You can change data and repeat the computation.

Example: An example of the static loading of a cylindrical shaft is built-in into the program. Select circular cross-section from the first page and input file BEAM1.DAT Hit ENTER repeatedly and the example will be executed. The input file is:

```
Problem Identification     beam problem?
Number of Elements:            3 ?
Modulus of Elasticity      210000000000 ?
Type of section,<1> Circular, <2> General 1?
     Element data                          node data
     length     Diameter     distr load    force        spring
1  ? 1            ? .1        ?             ?            ?
2  ? 1.2          ? .12       ? 300         ? 3000       ?
3  ? .8           ? .07       ? 400         ? 3000       ? 100000
   right end....                           ?            ?

Enter no of solid supports 2 ?
Enter numbers of support nodes:

Support  1   at node  1 ?
Support  2   at node  4 ?

Are the data correct (Y/N)  ? █
```

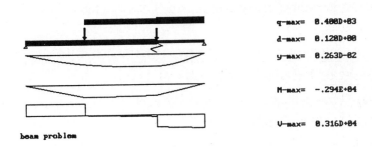

```
q-max=  0.400D+03

d-max=  0.120D+00

y-max=  0.263D-02

M-max=  -.294E+04

V-max=  0.316D+04
```

beam problem

```
Number of Nodes, Young's Modulus, type of section, problem ID
 3 , 2.1E+11 ,1,beam problem
Beam Data: length, section, distributed load, force, spring
 1 ,  .1 ,    0 ,    0 ,       0
 1.2 , .12 , 300 , 3000 ,       0
 .8 ,  .07 , 400 , 3000 , 100000
Rigid supports
 2
At the following nodes:
```

```
1 , 4
```

The results are plotted on the screen in the form of displacement, shear and bending moment diagrams, together with the beam, the supports and the loading. The results screen for BEAM1 is shown above.

The output file BEAMSTAT.RES is:

```
Number of Nodes, Young's Modulus, type of section, problem ID
 3 , 210000000000 ,1,beam problem
Beam Data: length, section, distributed load, force, spring
 1 , .1 ,   0 ,    0 , 0
1.2 , .12 , 300 , 3000 , 0
.8 , .07 , 400 , 3000 , 100000
Rigid supports
 2
At the following nodes:
 1 , 4
```

Node	Deflection	Slope	moment	Shear	Reaction
1	0.134D-06	0.263D-02	0.000D+00	-.294D+04	0.294D+04
2	0.215D-02	0.120D-02	-.294D+04	0.593D+02	0.000D+00
3	0.263D-02	-.388D-03	-.265D+04	0.316D+04	0.000D+00
4	0.158D-06	-.475D-02	0.767D-12	0.228D-07	0.348D+04

Probable Errors: Execution breaks: Hardware incompatibility, wrong graphics adaptor. Out of memory.

Deflections at supports large: Numerical problems, large D/L ratio for some element. Try changing units.

4.2. PREprocessing of FRAME geometry

Function: Creating 3-D frame models for structural analysis

Reference: Dimarogonas 1988, chapter 4.

Hardware Requirements: 512k, 1 FD 360k, EGA card, monochrome monitor.

Limitations: Up to 20 nodes, 20 elements. Limits can be extended by appropriate changes in the program.

Files needed: PREFRAME.EXE, COMPUME.SYS, data files FRAME*.DAT.

Method: Graphic interaction on 3 Cartesian views.

Use: To use the program (User types what below is underlined):

Select from COMPUME or type PREFRAME.

The first window is for program identification and basic data, maximum x and y or width and height of screen in the units to be used in the frame dimensions, or the maximum horizontal and vertical distances expected in the frame views. Enter xmax,ymax and hit the ENTER key.

Next is then the main menu window.

You are now in the main menu. You select with the keypad arrows. The > sign indicates the selection and at the bottom line there is an explanation of the command. ENTER invokes the respective command.

Quit quits the program execution.

Plot Frame plots a frame on screen.

New Starts a new frame. Any previous frame is erased.

Save File saves the current frame on disc. At prompt, enter file name desired (preferable with a *.fra extension). It prints a list of files with *.fra extension for reference. If no name is entered, you return to menu.

Load File loads a frame from disk. See above.

Clear clears the screen without erasing the frame from memory.

Make Model Creates the frame model. If you use this command, the Make Model menu appears on the screen:

N defines a node at the cross cursor which you can move with the direction keys of the keypad. The program will ask then for node data:

a. The constraints at the nodes. Up to six constraints can be specified along the respective coordinates.

b. The forces at the nodes. Up to six forces can be specified along the respective coordinates.

The program identifies node with a small circle, loading with a larger circle and constraint with a square. If you take the cursor in a previously defined node, the program prints loads or restraints already specified at this node. You can add or change data. If you do not specify new values, the program retains the old ones.

E creates an element between the last defined node or the end of the last defined element and the current position of the cross cursor. The program asks for element material and section properties. If many elements have the same properties, a number is assigned and if the same type of element is encountered the user enters only the element type number. The material properties will be specified in the FINFRAME program.

The following two commands are related to the motion of the cursor which moves with the keypad arrows.

F move the cursor in larger steps (multiplication by 10)

S move the cursor in smaller steps (division by 10)

L Locks cursor to the nearest node. It is used to define the beginning of an element to an already defined node. If the element ends on an already defined node, the program detects it if the position of the cursor when you hit E is within the cursor motion step in the x,y,z directions from a node.

M Exit to the main menu.

L To lock cursor in the nearest node, already defined.

X,Y,Z transfers control of the cursor on the respective screens.

P,D,C Polygon operations, see program solid. The elements created have the properties of the last one created.

Example: A portal frame is designed for structural analysis. The three-dimensional views of the frame and the data file saved FRAME1.DAT are shown below:

```
NODE,  110 ,  1600 ,  10
NODE,  110 ,  200  ,  10
NODE,  510 ,  200  ,  10
NODE,  510 ,  600  ,  10
NODE,  760 ,  600  ,  10
NODE,  1010 ,  600  ,  10
```

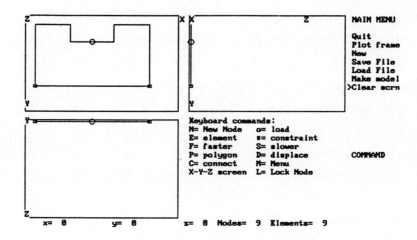

```
NODE, 1010 , 200 , 10
NODE, 1410 , 200 , 10
NODE, 1410 , 1600 , 10
ELEMENT , 1 , 1 , 2
ELEMENT , 2 , 2 , 3
ELEMENT , 3 , 3 , 4
ELEMENT , 4 , 4 , 5
ELEMENT , 5 , 5 , 6
ELEMENT , 6 , 6 , 7
ELEMENT , 7 , 7 , 8
ELEMENT , 8 , 8 , 9
ELEMENT , 9 , 9 , 1
PROPERTY , 1 , 1 , 1200 , 40000 , 360000 , 400000 , 0
PROPERTY , 2 , 1 , 1200 , 40000 , 360000 , 400000 , 0
PROPERTY , 3 , 1 , 1200 , 40000 , 360000 , 400000 , 0
PROPERTY , 4 , 1 , 1200 , 40000 , 360000 , 400000 , 0
PROPERTY , 5 , 1 , 1200 , 40000 , 360000 , 400000 , 0
PROPERTY , 6 , 1 , 1200 , 40000 , 360000 , 400000 , 0
PROPERTY , 7 , 1 , 1200 , 40000 , 360000 , 400000 , 0
PROPERTY , 8 , 1 , 1200 , 40000 , 360000 , 400000 , 0
PROPERTY , 9 , 1 , 1200 , 40000 , 360000 , 400000 , 0
LOADING @ , 5 , 0 ,-20000 , 0 , 0 , 0 , 0
RELEASE @ , 1 , 0 , 0 , 0 , 0 , 0 , 1
RELEASE @ , 9 , 1 , 0 , 0 , 0 , 0 , 1
FRAMEND
```

Probable Errors: Execution breaks: Very large mesh for the screen.

Hardware incompatibility, wrong graphics adaptor. Out of memory.

4.3. **FIN**ite element analysis of 3-D **FRAME**s

Function: Program PREFRAME must be used to create the frame or a file anyname.FRA prepared in the current directory before you use FINFRAME. You should know maximum x,y,z of frame. Finite Element Stress analysis of 3-D frames.

Reference: Dimarogonas 1988, chapter 4.

Hardware Requirements: 512k, 1 FD 360k, EGA card, monochrome monitor.

Limitations: Up to 20 nodes, 20 elements.

Files needed: FINFRAME.EXE, COMPUME.SYS, one data file FRAME*.DAT.

Method: For straight prismatic elements and constant properties EIalong the elements, end forces are related to end displacements with the element stiffness matrix, in a local coordinate system,

$$\underline{K}_{v(e)}\underline{x}_{(e)} = \underline{f}_{v(e)}$$

$$\underline{K}v(e) = EI \begin{bmatrix} 12/L^3 & & \text{symmetric} & \\ 6/L^2 & 4/L & & \\ -12/L^3 & -6/L^2 & 12/L^3 & \\ 6/L^2 & 2/L & 6/L^2 & 4/L \end{bmatrix}$$

where $x_{(e)}$ and $\underline{f}_v{}^{(e)}$ is the nodal displacement and force vectors.

Finally, the element stiffness matrices will be assembled to yield the system equations. The system equations are obtained directly:

$$\underline{K}q = \underline{f}_q \qquad\qquad\qquad (a)$$

where, after transformation to a global coordinate system,

$$\underline{K} = \sum_{i=1}^{n} \underline{K}_i{}^{(e)}$$

The displacements are obtained from the system of linear equations (a) with the Gauss Elimination method.

Use: Select from COMPUME <u>FINFRAME</u>. After the sign-in page, the program asks for the file where the file prepared with program PREFRAME. It has a name FRAMExxx. The program plots the frame first in an identical screen with the one in PREFRAME.

Then the program prints data, member forces and node displacements. It prints also forces in the local coordinate system, for stress analysis, at the two ends of the element. Finally, it plots the frame in the original and the displaced positions.

Example: The portal frame in file FRAME1.DAT is used as input. Steel properties $E = 2.1 \times 10^{11}$ and $G = 1.05 \times 10^{11}$ are entered. The program output, file FRAME1.RES and the screen plot of the undeflected and the deflected frame follow:

```
Output file from FINFRAME:
Input File Name: FRAME1.DAT
NODE DISPLACEMENTS
```

Node	ux	uy	uz	rotx	roty	rotz
1	0.000E+00	0.000E+00	0.000E+00	0.000E+00	0.000E+00	0.112E-08
2	-.688E-05	-.556E-07	0.000E+00	0.000E+00	0.000E+00	-.194E-07
3	-.688E-05	-.902E-05	0.000E+00	0.000E+00	0.000E+00	-.219E-07
4	-.715E-08	-.903E-05	0.000E+00	0.000E+00	0.000E+00	-.118E-07
5	-.521E-08	-.107E-04	0.000E+00	0.000E+00	0.000E+00	0.380E-12
6	-.327E-08	-.903E-05	0.000E+00	0.000E+00	0.000E+00	0.118E-07
7	0.687E-05	-.902E-05	0.000E+00	0.000E+00	0.000E+00	0.219E-07
8	0.687E-05	-.556E-07	0.000E+00	0.000E+00	0.000E+00	0.194E-07
9	-.101E-07	0.000E+00	0.000E+00	0.000E+00	0.000E+00	-.112E-08

ELEMENT FORCES-Global Coordinate system

Element	Fx	Fy	Fz	Mx	My	Mz
1	-1.95E+03	1.00E+04	0.00E+00	0.00E+00	0.00E+00	-2.60E+05
2	-1.95E+03	1.00E+04	0.00E+00	0.00E+00	0.00E+00	2.47E+06
3	-1.95E+03	1.00E+04	0.00E+00	0.00E+00	0.00E+00	-1.53E+06
4	-1.95E+03	1.00E+04	0.00E+00	0.00E+00	0.00E+00	-2.31E+06
5	-1.95E+03	-1.00E+04	0.00E+00	0.00E+00	0.00E+00	-4.81E+06
6	-1.95E+03	-1.00E+04	0.00E+00	0.00E+00	0.00E+00	-2.31E+06
7	-1.95E+03	-1.00E+04	0.00E+00	0.00E+00	0.00E+00	-1.53E+06
8	-1.95E+03	-1.00E+04	0.00E+00	0.00E+00	0.00E+00	2.47E+06
9	-1.95E+03	7.72E-03	0.00E+00	0.00E+00	0.00E+00	-1.30E+05

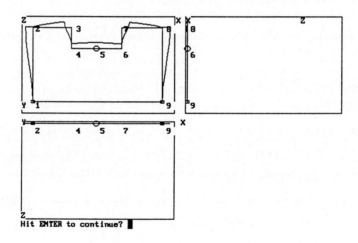

Hit ENTER to continue?

```
ELEMENT FORCES-Local Coordinate system
Element       Fx        Fy        Fz        Mx        My        Mz

 1  (left:) -1.00E+04 -1.95E+03  0.00E+00  0.00E+00  0.00E+00 -2.60E+05
    (right:) 1.00E+04  1.95E+03  0.00E+00  0.00E+00  0.00E+00 -2.47E+06
 2  (left:) -1.95E+03  1.00E+04  0.00E+00  0.00E+00  0.00E+00  2.47E+06
    (right:) 1.95E+03 -1.00E+04  0.00E+00  0.00E+00  0.00E+00  1.53E+06
 3  (left:)  1.00E+04  1.95E+03  0.00E+00  0.00E+00  0.00E+00 -1.53E+06
    (right:)-1.00E+04 -1.95E+03  0.00E+00  0.00E+00  0.00E+00  2.31E+06
 4  (left:) -1.95E+03  1.00E+04  0.00E+00  0.00E+00  0.00E+00 -2.31E+06
    (right:) 1.95E+03 -1.00E+04  0.00E+00  0.00E+00  0.00E+00  4.81E+06
 5  (left:) -1.95E+03 -1.00E+04  0.00E+00  0.00E+00  0.00E+00 -4.81E+06
    (right:) 1.95E+03  1.00E+04  0.00E+00  0.00E+00  0.00E+00  2.31E+06
 6  (left:)  1.00E+04 -1.95E+03  0.00E+00  0.00E+00  0.00E+00 -2.31E+06
    (right:)-1.00E+04  1.95E+03  0.00E+00  0.00E+00  0.00E+00  1.53E+06
 7  (left:) -1.95E+03 -1.00E+04  0.00E+00  0.00E+00  0.00E+00 -1.53E+06
    (right:) 1.95E+03  1.00E+04  0.00E+00  0.00E+00  0.00E+00 -2.47E+06
 8  (left:) -1.00E+04  1.95E+03  0.00E+00  0.00E+00  0.00E+00  2.47E+06
    (right:) 1.00E+04 -1.95E+03  0.00E+00  0.00E+00  0.00E+00  2.60E+05
 9  (left:)  1.95E+03  7.72E-03  0.00E+00  0.00E+00  0.00E+00  1.30E+05
    (right:)-1.95E+03 -7.72E-03  0.00E+00  0.00E+00  0.00E+00 -1.30E+05
```

Probable Errors: Execution breaks: Out of memory. Division by zero or overflow: Not enough constraints. The model has rigid body motion.

4.4. FINite element STRESs analysis

Function: Finite Element Stress analysis of plane or axisymmetric solids.

Files needed: Program AUTOMESH must be used to create the triangular mesh. A file MESH*.DAT must exist in the current directory before you use FINSTRES.

Reference: Dimarogonas 1988, chapter 4.

Hardware Requirements: 512k, 1 FD 360k, EGA card, monochrome monitor.

Limitations: Up to 30 nodes, 30 elements. Limits can be extended by appropriate changes in the program.

Files needed: FINSTRES.EXE, COMPUME.SYS, one MESH*.DAT file.

Method: For triangular plane elements with displacements in the plane of element, the nodal displacements $\underline{x}^{(e)} = \{u_1\ v_1\ u_2\ v_2\ u_3\ v_3\}$ are related with the corresponding nodal forces $\underline{f}_v^{(e)}$ as

$$\underline{K}_v^{(e)}\underline{x}_{(e)} = \underline{f}_v^{(e)}$$

where $\underline{K}_v^{(e)}$ is the element stiffness matrix. Finally, the element stiffness matrices will be assembled to yield the system equations. The system equations are obtained directly:

$$\underline{K}\underline{q} = \underline{f}_q$$

where

$$\underline{K} = \sum_{i=1}^{n} \underline{K}_i{}^{(e)}$$

The displacements are obtained from the system of linear equations (a) with the Gauss Elimination method.

Use: To use the compiled program, select from COMPUME or type <u>FINSTRES</u>.

The first window is for program identification and basic data:

a. Type of problem:

a1. Plane stress. A thin plate with in-plane loads.

a2. Plane strain. A thick plate with in-plane loads.

a3. Axisymmetric, Plane stress. A thin cylindrical disc with axisymmetric, axial or radial loads.

a4. Axisymmetric, Plane strain. A general axisymmetric solid with axisymmetric, axial or radial loads.

b. Material properties: Modulus of Elasticity, Poisson Ratio, thickness for the first two problem types. Use a consistent system of units.

Next is then the main menu window.

You are now in the main menu. You select with the keypad direction keys. The > sign indicates the selection and at the bottom line there is an explanation of the command. ENTER invokes the respective command.

Quit: quits the program execution.

Load file: loads a mesh from disk. The mesh was generated with the AUTOMESH program.

Plot Mesh: Plots the mesh currently in memory. Caution: For plane problems, the mesh can be anywhere in the screen. For axisymmetric problems, the y-axis must coincide with the axis of symmetry. Only half of the section must be in the mesh. The program assumes that radii are measured from the y-axis.

Loads: To specify the node forces.

Restraints: To specify node restraints. Both commands have the same menu, one can work with either:

N finds the nearest node to the cross cursor. This cursor moves with the keypad direction keys.

The following two commands are related to the motion of the cross cursor:

F move the cursor in larger steps (multiplication by 10)

S move the cursor in smaller steps (division by 10)

M Exit to the main menu.

L specify load at current node. Move cursor near the desired node and hit N. The cursor locks on the node. Hit L. The program asks for Fx and Fy. These are the loads in the x and y direction, respectively, for the first two types of problems. For axisymmetric problems, they are, total peripheral radial and axial node load, respectively. The program asks also for body forces. In the plane problems, these are the acceleration of gravity in compatible units. For the axisymmetric cases the y body force can be gravity, if the axis of symmetry is the vertical one physically. The x body force is density x ω^2, the radial force due to the rotation with angular velocity ω about the geometric axis.

R: You specify restraints at the current node. Move cursor near the desired node and hit N. The cursor locks on the node. Hit R. The program asks for restraint in the x, y directions. Specify one or two or both. Make sure that you have enough restraints to avoid rigid body motion.

Analyze: The program performs the analysis and prints nodal forces and displacements.

Post Proc: Plots the original and displaced mesh and the stress distribution in a color code.

EXAMPLE: An arched bridge was divided into finite elements with the AUTOMESH program above. The MESH2.DAT file will be used here. Supports will be defined on the lower two corners and a load at the mid-span on the top. The screen plots of the mesh with the load and supports and the color map of the stresses follow. The results are in file FINSTRES.RES.

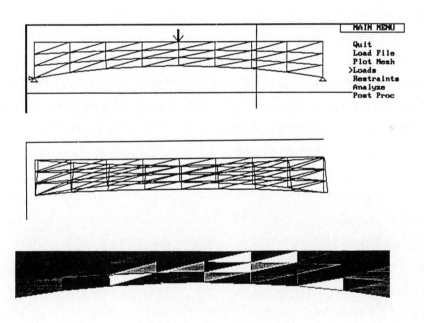

Probable Errors: Execution breaks: Very large mesh for the screen. Hardware incompatibility, wrong graphics adaptor. Out of memory. Execution breaks at post processing: Missing EGA card. Division by zero or overflow: No mesh loaded. Not enough constraints. The model has rigid body motion.

4.5. Transfer Matrix stability analysis of COLUMNs

Function: Stability Analysis of Beam-columns, Transfer Matrix Method.

Reference: Dimarogonas 1988, chapter 9.

Hardware Requirements: 512k, 1 FD 360k, EGA card, monochrome monitor.

Limitations: Up to 10 nodes, 9 elements, 3 solid supports.

Files needed: TMCOLUMN.EXE, COMPUME.SYS, data files [COLUMN*.DAT].

Method: The transfer matrix method is used.

Use: To use the program (User types what below is underlined):

Select from COMPUME or type TMCOLUMN.

The first window is for program identification. Define the type of section (circular or general) and two files: one to read data from, the other to write the data to, if you edit them. If you do not want to deal with files, enter nothing (just hit ENTER).

Next is then the data window.

Enter number of elements, modulus of elasticity at the ? prompt and ENTER.

Every subsequent line corresponds to a node or element numbered consecutively from left to right. Every time you hit ENTER, the cursor moves to the next entry. To enter a new value, type it and hit ENTER. For zero or to maintain the previous value, just hit ENTER. The first 2 entries are for the element:

Length of the element, distance of node from last one.

Minimum area moment of inertia for a general section,

Diameter if you have selected circular section.

Nominal axial load on the element

The next 2 entries in the line are for node data:

Lateral force on the node

Spring constant of flexible support, if such support

exists. Do not include here solid supports.

If you enter zero length or section, the line is repeated.

The last line has node data only because there is one node more than elements.

Then the program asks for the number of solid supports. Enter their number and node numbers of their location, as the program requests them.

Finally, the program asks if data are correct. If the answer is no, the program returns in the beginning. The data entered are printed on their position. Type only the corrections or additions. For all correct entries, just hit ENTER. You can repeat this as many times as you like.

If the data are correct, the program changes page and asks for the axial load, minimum, maximum value and step for the search for the maximum additional load. To assist you, the program prints two extreme values for the column simply supported et the ends and having uniform section, the maximum and minimum section respectively.

Then it proceeds with the analysis. It plots the value of the characteristic determinant and the critical load, if found.

If you respond R to the question "Hit ENTER to continue ?", the program returns to the data menu. You can change any entry and repeat the computation.

Example: The critical load for a compound column will be found with TMCOLUMN. Circular cross-section is selected and the input file COLUMN1.DAT is prepared:

```
3,2.1E+11,"1","test column"
1  , .02,1000,0   ,0
1.5, .03,2000,1000,0
1  , .02,3000,0   ,0
2
1,4
```

The data preparation and the results screen are shown below:

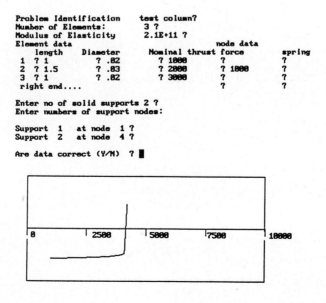

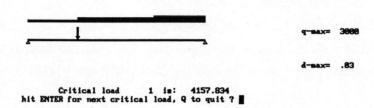

q-max=   3000

d-max=   .83

Critical load   1 is: 4157.834
hit ENTER for next critical load, Q to quit ? █

Probable Errors: Overflow: Elements have short length and single precision is not sufficient.

No critical load is found: Upper search limit too low or range too narrow or step too high.

5

CHAPTER FIVE
ANALYSIS OF MECHANISMS

5.1. SIMUlation of MECHanisms.[*]

Function: Kinematic and dynamic simulation of planar linkages.

Reference: Nikravesh 1988.

Hardware Requirements: 512k, HD, EGA card, color monitor.

Limitations:

Method: The mechanism is modeled by a number of solids connected by rotational and translational joints. Mechanism geometry and motion is confined to a single plane. For every joint, constraint equations are written for the motion of the solids it connects.

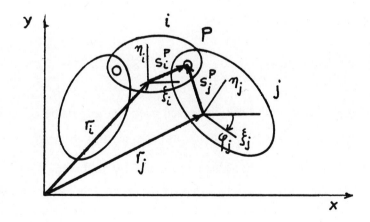

---

* Contributed by Dr. T. Chondros, Department of Mechanical Engineering, University of Patras, Greece.

Let two solids, i and j, be connected by a rotational joint P, (x,y) a fixed coordinate system, $\underline{r}_i$ and $\underline{r}_j$ the position vectors of two points O_i and O_j on solids i and j, respectively, on the global coordinate system and $\underline{s}_i^P$, $\underline{s}_j^P$ the vectors connecting points O_i and O_j to the point P of the joint. The constraint equation is

$$\underline{r}_i + \underline{s}_i^P - \underline{r}_j - \underline{s}_j^P = 0 \qquad\qquad (a)$$

The vectors $\underline{s}_i^P$ and $\underline{s}_j^P$ have fixed length and variable orientation, they rotate with the respective solids i and j. Therefore, if $\underline{s}_{0i}^P$ and $\underline{s}_{0j}^P$ are the values of these vectors at the initial position of the mechanism and ϕ_i, ϕ_j the angles of rotation of the solids i and j, the constraint equation becomes

$$\underline{r}_i + \underline{s}_{0i}^P e^{i\phi_i} - \underline{r}_j - \underline{s}_{0j}^P e^{i\phi_j} = 0 \qquad\qquad (b)$$

This vector equation results in two algebraic equations. The scalar unknowns are six, the two vectors $\underline{r}_i$ and $\underline{r}_j$ and the two angles ϕ_i and ϕ_j. Therefore the system has four degrees of freedom. Each revolute joint reduced the number of degrees of freedom in a planar mechanism by 2.

A four-bar linkage has a grounded link and three moving links with four revolute joints. The degrees of freedom are 3x3 - 2x4 = 1. Therefore, the scalar constraint equations are 8 and the scalar unknowns are 9. One unknown is undetermined. If, however, the angle of rotation of one link, the input link, is known then there are 8 equations in 8 unknowns and they can be solved to find the position of the mechanism at any rotation of the input link.

Similar reasoning can be used for translational, geared and other joints. It will result, in general, in a system of nonlinear algebraic equations

$$f_i[x_1, x_2, \ldots, x_n; \phi(t)] = 0, \quad i = 1, 2, \ldots, n \qquad\qquad (c)$$

where $\phi(t)$ is the input which varies with time in a given way. For any value of time, the unknowns x_1, x_2, ..., x_n should be determined. In the case of a four-bar linkage, input is the angle of rotation of the crank and the two unknowns are the angles of rotation of the coupler and the follower.

Position analysis, finding the output for any value of the input, can be preformed numerically with the Newton algorithm. Starting from the initial position $\phi = \phi_0$, the input is increased by an incremental value $\Delta\phi$. A Taylor expansion of the functions in equations (c) about the initial position of the mechanism, retaining only linear terms, yields

$$f_i[x_1, x_2, \ldots, x_n; \phi_0 + \Delta\phi] = f_i[x_{01}, x_{02}, \ldots, x_{0n}; \phi_0] +$$

$$+(\partial f_i/\partial x_1)\dot{\Delta}x_1 + (\partial f_i/\partial x_2)\Delta x_2 + \ldots + (\partial f_i/\partial x_n)\Delta x_n + (\partial f_i/\partial\phi)\Delta\phi \qquad (d)$$

Since f_i are always 0 due to equation (c), equation (d) becomes

$$(\partial f_i/\partial x_1)\Delta x_1 + (\partial f_i/\partial x_2)\Delta x_2 + \ldots + (\partial f_i/\partial x_n)\Delta x_n = -(\partial f_i/\partial\phi)\Delta\phi \qquad (e)$$

Equations (e) can be written in matrix form

$$\underline{\Phi}\underline{y} = \underline{b} \qquad (f)$$

where $\underline{\Phi} = [\partial f_i/\partial x_j]$ is the Jacobian of the system, $\underline{y} = \{\Delta x_1, \Delta x_2, \ldots, \Delta x_n\}$ is the vector of the increments of the unknowns and $\underline{b} = \{\partial f_i/\partial\phi\}$ a known constant vector. The increments of the unknowns can be computed from the system of linear equations (f) and the solution can be constructed step-by-step for a given sequence of input values ϕ_0, $\phi_0 + \Delta\phi$, $\phi_0 + 2\Delta\phi$, $\phi_0 + 3\Delta\phi$, ...

Velocity analysis, finding the output velocity for any value of the input velocity, can be preformed numerically with the Newton algorithm. Starting from the initial position $\phi = \phi_0$, the time is increased by an incremental value Δt. A Taylor expansion of the functions in equations (c) about the initial position of the mechanism, retaining only linear terms, yields

$$f_i[x_1, x_2, \ldots, x_n; \phi(t + \Delta t)] = f_i[x_{01}, x_{02}, \ldots, x_{0n}; \phi_0] +$$

$$+(\partial f_i/\partial x_1)x_1{}^\circ\Delta t + (\partial f_i/\partial x_2)x_2{}^\circ\Delta t + \ldots + (\partial f_i/\partial x_n)x_n{}^\circ\Delta t + (\partial f_i/\partial\phi)\phi^\circ\Delta t \qquad (g)$$

Since f_i are always 0 due to equation (c), equation (g) becomes

$$(\partial f_i/\partial x_1)x_1{}^\circ + (\partial f_i/\partial x_2)x_2{}^\circ + \ldots + (\partial f_i/\partial x_n)x_n{}^\circ = -(\partial f_i/\partial\phi)\phi^\circ \qquad (h)$$

Equations (h) can be written in matrix form

$$\underline{\Phi}\underline{v} = \underline{b}\phi^\circ \qquad (i)$$

where $\underline{\Phi} = [\partial f_i/\partial x_j]$ is the Jacobian of the system, $\underline{v} = \{x_1{}^\circ, x_2{}^\circ, \ldots, x_n{}^\circ\}$ is the vector of the unknown velocities. The velocities can be computed from the system of linear equations (f) and the solution can be constructed step-by-step for a given sequence of input values ϕ_0, $\phi_0 + \Delta\phi$, $\phi_0 + 2\Delta\phi$, $\phi_0 + 3\Delta\phi$, ...

Acceleration analysis can be performed by further differentiation of equations (h)

$$(\partial f_i/\partial x_1)x_1{}^{\circ\circ} + (\partial f_i/\partial x_2)x_2{}^{\circ\circ} + \ldots + (\partial f_i/\partial x_n)x_n{}^{\circ\circ} =$$

$$-[(\partial_2 f_i/\partial x_1{}^2)x_1{}^\circ + (\partial_2 f_i/\partial x_2{}^2)x_2{}^\circ + \ldots + (\partial_2 f_i/\partial x_n{}^2)x_n{}^\circ$$

$$-(\partial f_i/\partial\phi)\phi^{\circ\circ} - (\partial_2 f_i/\partial\phi^2)\phi^\circ \qquad (j)$$

In matrix form

$$\underline{\Phi}\underline{a} = -\underline{H}\underline{v} + \underline{b}^\circ + \underline{f}\phi^\circ \qquad (k)$$

where $\underline{H} = [\partial_2 f_i/\partial x_j{}^2]$ is the Hessian matrix of the system, $\underline{f} = \{\partial_2 f_i/\partial\phi^2\}$ and $\underline{a} = \{x_1{}^{\circ\circ}, x_2{}^{\circ\circ}, \ldots, x_n{}^{\circ\circ}\}$ is the vector of the unknown acceleration. The accelerations can be computed from the system of linear equations (k) and the solution can be constructed step-by-step for a given sequence of input values ϕ_0, $\phi_0 + \Delta\phi$, $\phi_0 + 2\Delta\phi$, $\phi_0 + 3\Delta\phi$, ...

The accuracy of the method depends on the accuracy of the solution of the position problem. Sufficiently small step Δt must be used to reduce the truncation error, observing also the round-off error for very small Δt.

Use: This module performs kinematic and dynamic analysis of mechanical systems. It performs six basic functions:

1. Configuring the system
2. Accepting data from the user
3. Generating the governing equations of motion
4. Solving the equations
5. Communicating the results to the user
6. Saving the results

The pull-down menu is activated with the selection of SIMUMECH from the root menu and the desired application may be selected.

The program produces a library with all the mechanisms being analyzed. This option provides the user with a graphical display of various mechanisms which is helpful in the design process. Input is provided to both the Kinematic Analysis and the Dynamic Analysis modules through the console via the <Data Entry> option. Any text editor may be also used to create a data entry file. The output is written on the console screen with a specified format. Output data are stored automatically in a data file during the data entry process.

A Cartesian coordinate system is used to specify the position of all bodies of a mechanism. In order to specify the configuration of a planar system, a body fixed (η_i, ζ_i) coordinate system is embedded in each body of the system as shown in the Figure below. Each body can be located by specifying the global translational coordinates of the origin of the body-fixed (η_i, ζ_i) reference system and the angle of rotation of this system relative to the global (x, y) axes. This angle is considered positive if the rotation from positive x-axis to positive ξ_i-axis is counterclockwise.

Body-fixed coordinates (η_i, ζ_i) are attached to each body, including the ground. Positioning of these coordinate systems is quite arbitrary for kinematic analysis. However, it is a good practice to locate the origin of the coordinate system to the center of gravity of the body. Furthermore, aligning at least one of the coordinate axes with the link axis or parallel to some line of certain geometric or kinematic importance may simplify the task of collecting data for the kinematic pairs in the system.

The revolute joints are defined by their coordinates relative to the local coordinate (η_i, ζ_i) systems of the bodies they link. A revolute-revolute joint is defined as a link of length L which connects two revolute joints in the two solids i, j it connects.

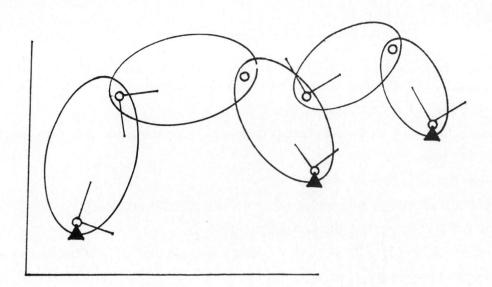

For the translational joint, two points on body i and one point on body j are selected on a line parallel to the line of translation. Similarly, for a revolute-translational joint between two bodies i and j three points have to be defined. The first is the revolute joint on body i that can move along the line of translation on body j, and two arbitrary points on the line of translation which are defined with respect to body j local coordinate system. Although similar procedures are followed for the Dynamic analysis.

The input prompts that the user has to answer for kinematic analysis are:

a. ENTER NB, NR, NT, NRR, NTR, NG, NS, ND, NP?

where,

       NB = Number of bodies in the system, including ground

       NR = Number of revolute joints in the system

       NT = Number of translational joints in the system

       NRR = Number of revolute-revolute joints in the system

       NTR = Number of revolute-translational joints in the system

       NG = Number of bodies that are attached to (or considered as) ground

       NS = Number of simple constraints in the system

       ND = Number of driving constraints (driving links)

       NP = Number of points of interest

The program computes the number of coordinates N and the total number of constraint equations M, including the driving constraints, from this first set of data. If N is not equal to M, then an error message is printed.

b. FOR BODY k ENTER INITIAL ESTIMATES ON X,Y,PHI.

From an initial sketch of the system, the user should measure <u>approximate</u> values for the coordinates of the local coordinate system in respect to the global one and the angle from the x-axis to the ξ-axis, ccw

  c. FOR REVOLUTE JOINT k ENTER NUMBERS OF BODIES I AND J, LOCAL COORDINATES

     Ksi_P_I, Eta_P_I, Ksi_P_J, Eta_P_J

The location of the joint on the two bodies it connects is given in the two local coordinate systems,

     $i, j, \xi_{Pi}, \eta_{Pi}, \xi_{Pj}, \eta_{Pj}$

  d. FOR TRANSLATIONAL JOINT k ENTER FOR BODY NUMBERS I AND J

     Ksi_P_I, Eta_P_I, Ksi-Q-I, Eta-Q-I, Ksi_P_J, Eta_P_J

The location of the joint P on the body j and point P and Q if the joint slides along the line PQ on the body i, given in the two local coordinate systems,

     $i, j, \xi_{Pi}, \eta_{Pi}, \xi_{Qi}, \eta_{Qi}, \xi_{Pj}, \eta_{Pj}$

  e. FOR REVOLUTE-REVOLUTE JOINT NO. k ENTER BODY NUMBERS I,

     Ksi_P_I, Eta_P_I, Ksi_Q_J, Eta_Q_J, LENGTH

The location of the joints P and Q on the two bodies is given in the two local coordinate systems, and the length of the connecting link PQ,

     $i, j, \xi_{Pi}, \eta_{Pi}, \xi_{Qj}, \eta_{Qj}, L$

  f. FOR REVOLUTE-TRANSLATIONAL JOINT NO. k ENTER BODY NUMBERS I AND J,

     Ksi_P_I, Eta_P_I, Ksi-Q-J, Eta-Q-J, Ksi-E-J, Eta-E-J

The location of the joint P on the body i and point Q and E if the joint slides along the line QE on the body j, given in the two local coordinate systems,

     $i, j, \xi_{Pi}, \eta_{Pi}, \xi_{Qi}, \eta_{Qi}, \xi_E, \eta_E$

  g. ENTER BODY NUMBER FOR THE GROUNDED BODY NUMBER k.

  h. FOR SIMPLE CONSTRAINT NO. k ENTER BODY NO.

     AND 1,2, OR 3 FOR X,Y,OR PHI CONSTRAINT DIRECTION

  i. FOR DRIVER NUMBER k ENTER BODY NUMBER, THE DIRECTION OF INPUT MOTION

     1,2, OR 3 FOR X,Y, OR PHI, INITIAL POSITION, VELOCITY, AND ACCELERATION

  j. FOR POINT OF INTEREST NO k ENTER BODY NUMBER, Ksi-P AND Eta-P COORDINATES

The location of any reference point, for which the motion needs to be computed, for example,

     i, ξ_{Pi}, η_{Pi}

   k. ENTER STARTING TIME, FINAL TIME, AND TIME INCREMENT

     $t_0, t_{max}, \Delta t$

For dynamic analysis, the program requests the number of translational spring, damper or actuator elements. An element can have one spring, one damper, and one actuator as long as the attachment points are shared. The local coordinates of the connecting points to bodies i, j as well as the spring constant k, the damping coefficient d, the actuator force f, and the undeformed spring length are introduced for each element.

The forces acting on member i are stored as fx, fy, n where n is the applied moment. The program calculates and reports the reactions at the kinematic pairs.

Input Prompts :

a. ENTER NB, NR, NT, NG, NS, NSP, NP.

   NB = Number of bodies in the system, including ground

   NR = Number of revolute joints in the system

   NT = Number of translational joints in the system

   NG = Number of bodies that are attached to (or considered as) ground

   NS = Number of simple constraints in the system

   NSP = Number of translational spring, damper, or actuator elements

   NP = Number of points of interest

b. FOR BODY k ENTER INITIAL DISPLACEMENTS X, Y, PHI,

      INITIAL VELOCITIES XD, YD, PHID,

      MASS, MOMENT OF INERTIA,

      CONSTANT FORCE-MOMENT FX, FY, M

c. FOR REVOLUTE JOINT NO.k ENTER BODY NOS.I AND J

      Ksi_P_I, Eta_P_I, Ksi_P_J, Eta_P_J

d. FOR TRANSLATIONAL JOINT NO. k ENTER BODY NOS. I AND J

      Ksi_P_I, Eta_P_I, Ksi-Q-I, Eta-Q-I, Ksi_P_J, Eta_P_J

e. ENTER BODY NO. FOR NO. k GROUNDED BODY

f. FOR SIMPLE CONSTRAINT NO. k ENTER BODY NO.

      AND 1,2, OR 3 FOR X,Y,OR PHI CONSTRAINT DIRECTION

g. FOR SPRING ELEMENT NO. k ENTER BODY NOS I AND J   ,

      Ksi_P_I, Eta_P_I, Ksi_P_J, Eta_P_J,

      SPRING CONSTANT, DAMPING COEFFICIENT, ACTUATOR FORCE,

      UNDEFORMED SPRING LENGTH

h. FOR POINT OF INTEREST NO k ENTER BODY NO,

      Ksi-P AND Eta-P COORDINATES,

i. ENTER STARTING TIME, FINAL TIME, AND TIME INCREMENT

Example I: For the default problem of a four bar linkage shown (From Nikravesh 1988, by permission) the following data are given as in file <SIMUMEC1.DAT>:

SIMUMEC1.DAT

a	NB,NR,NT,NRR,NRT,NG,NS,ND,NP	4,4,0,0,0,1,0,1,1
b	ground body initial estimates on x,y,phi	0,0,0
b	body no 2 initial estimates on x,y,phi	5, .8,1.047
b	body no 3 initial estimates on x,y,phi	2.6,2.6, .5
b	body no 4 initial estimates on x,y,phi	3.5,1.8,1.047
c	data for rev. joint no 1	1 , 2, 0, 0, -1, 0
c	data for rev. joint no 2	2 , 3, 1, 0, -2, 0
c	data for rev. joint no 3	3 , 4, 2, 0, 2, 0
c	data for rev. joint no 4	4 , 1,-2, 0,2.5, 0

e	number of grounded body	1
i	data for driver no 1	2,3,1.047,6.2832,0
j	data for point of interest	3,.5,1.5
k	starting, final time, time step	0, 1, .025

The Kinematic Analysis option is used from the SIMUMECH menu. The program computes the displacement, velocity and acceleration and prints them on the screen and into the results file. Moreover, it produces a file EXAMPLE1.PLO which can be used subsequently with the Plotting selection to plot and animate the mechanism.

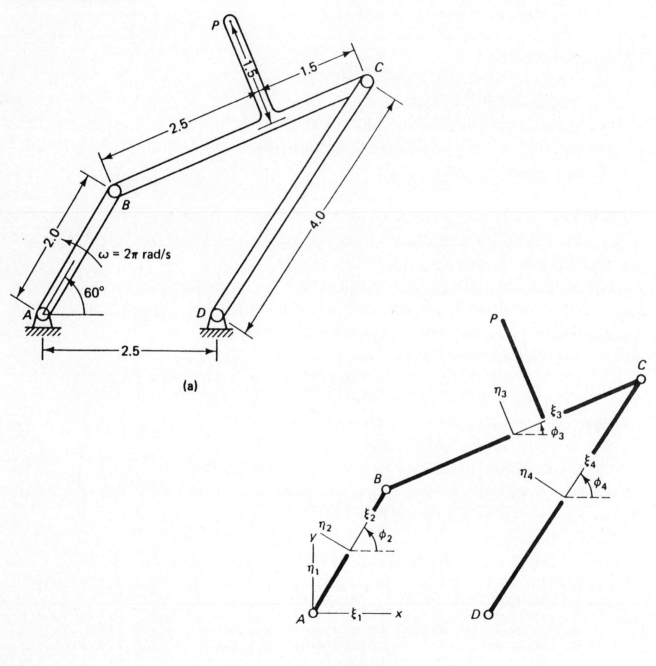

(a)

(b)

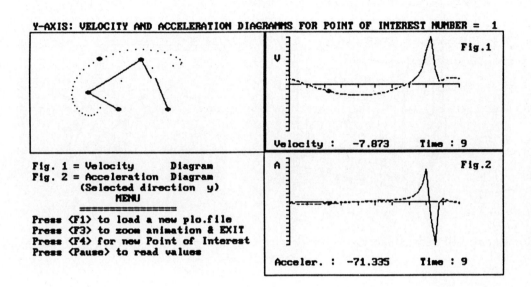

Example II: The four-bar linkage of the previous example is considered again. The mechanism is released from an initial position and falls under its own weight. Mass and moments of inertia for the moving bodies are given as in data input while for body 1, the non-moving body, any arbitrary value can be assigned (From Nikravesh 1988, by permission).

For the default problem of a four bar linkage (DYNAM1.DAT) input data are:

```
                                                DYNAM1.DAT
a    NB,NR,NT,NG,NS,NSP,NP              4,4,0,1,0,0,1
b    ground   initial conditions       0,0,0,0,0,0,0,0,0,0,0
b    body no 2 initial conditions       5,.8,1.047,0,0,0,1,.3,0,0,0
b    body no 3 initial conditions       2.8,2.5,.4,0,0,0,2.2,2,0,0,0
b    body no 4 initial conditions       3.5,1.7,1.0,0,0,0,2,1.3,0,0,0
c    data   for rev. joint no 1         1 , 2, 0, 0, -1, 0
c    data   for rev. joint no 2         2 , 3, 1, 0, -2, 0
c    data   for rev. joint no 3         3 , 4, 2, 0, 2, 0
c    data   for rev. joint no 4         4 , 1,-2, 0, 2.5, 0
e    number of grounded body        1
h    data for point of interest       3, .5, 1.5
i    starting, final time, time step 0, 1, .025
```

The Dynamic Analysis option is used from the SIMUMECH menu. The procedure is similar with the one for kinematic analysis.

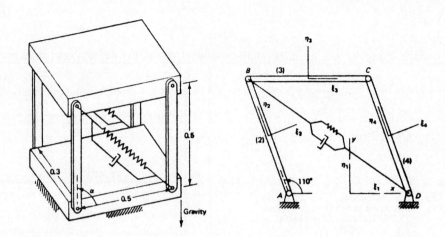

Probable Errors: Ill-defined mechanism, inconsistent data, wrong time increment Δt, mechanism stalls.

6

CHAPTER SIX
VIBRATION ANALYSIS
SIMUlated vibration LABoratory

6.1. PENDULUM.

Function: Simulation of the pendulum operation. The period of oscillation of the pendulum can be found as function of the maximum amplitude of the oscillation. It is known that this period is constant for small amplitudes but for larger amplitudes the period becomes larger and tends to infinity as the maximum oscillation amplitude tends to 180°.

Limitations: None.

Files Needed: PENDULUM.EXE, COMPUME.SYS

Method: Numerical solution of the pendulum equation

$$\theta" + (g/L)\sin\theta = 0$$

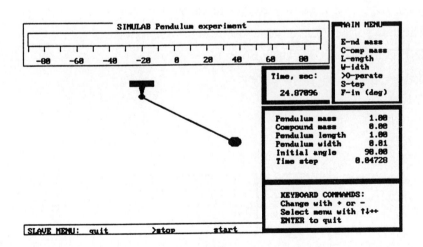

in the normal form

$$\theta' = \omega$$

$$\omega' = -(g/L)\sin\theta$$

The initial conditions are, initial angular displacement $\theta = \theta_0$, initial angular velocity $\theta' = 0$.

Use: On the right of the screen there is the main menu and on the screen bottom is the slave menu, submenu of the main menu. Selection is done with the keypad arrows or by hitting the initial letter for the main menu commands.

To change the system parameters: Select <E-nd mass>, <C-ompound mass>, <Length>, <Width>, initial angular displacement <F-in> from the MAIN MENU. Use + to increase the parameter by one step, use - to decrease the parameter by one step select <S-tep> in the main menu to change the step. Use + to increase the step, use - to decrease the step. Observe values of the parameters in the parameter widow, on the right. Select MAIN and SLAVE menus with keypad arrows.

To operate: Select <O-perate> from the main menu and <Quit>, <Stop>, <Start> from the slave menu. Observe the motion on the screen. The angular displacement is shown on the linear scale on the top of the screen.

Example: Find the relation of the natural frequency of a simple pendulum to the maximum oscillation amplitude (Dimarogonas & Haddad 1992, page 28).

PENDULUM is selected from the SIMULAB menu. The laboratory screen follows. By selecting initial angles from 20 to 180 degrees and measuring the time of the pendulum T_{20} to perform 20 oscillations, the natural period of oscillation is $T_n = T_{20}/20$. It is plotted as function of the initial angle. The parameters used are shown on the screen.

Probable Errors: Absence of the COMPUME.SYS file, wrong system configuration.

6.2. **LINEar SYStem analysis**

Function: Simulation of a 2-degree of freedom linear system for natural and forced vibration. The forcing mechanism can be harmonic or periodic or general forces or base motions.

Files Needed: LINESYS.EXE, COMPUME.SYS

Method: A two degree of freedom linear system

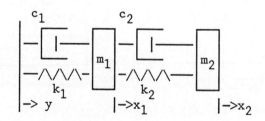

is modeled with the system of equations (Dimarogonas & Haddad 1992):

$$m_1 x_1'' + (c_1 + c_2)x'_1 - c_2 x'_2 + (k_1 + k_2)x_1 - k_2 x_2 = k_1 y_1 + k_2 y'_1 +$$
$$+ F_{11}\cos(\omega t + \phi_{11}) + F_{12}\cos(2\omega t + \phi_{12}) + F_{13}\cos(3\omega t + \phi_{13}) + \ldots$$
$$m_2 x_2'' - k_2 x_1 + k_2 x_2 - c_2 x'_1 + c_2 x'_2 =$$
$$= F_{21}\cos(\omega t + \phi_{21}) + F_{22}\cos(2\omega t + \phi_{22}) + F_{23}\cos(3\omega t + \phi_{23}) + \ldots$$

with initial conditions

$$x_1(0) = x_{10}, \; x_2(0) = x_{20}, \; x_1'(0) = v_{10}, \; x_1'(0) = v_{10}$$

The equations are solved by numerical integration.

Use: On the title page, the program prompts for a single mass or a two-mass system. Then, the main screen appears. On the right there is the main menu and on the screen bottom is the slave menu, submenu of the main menu. Selection is done with the keypad arrows. The program has on-line help.

To change the scale: Select SCALE in the main menu, select AMPLITUDE or TIME in the slave menu. Use - to scale down (smaller amplitude or shorter period), use + to scale up (larger amplitude or period). Observe the new scale in the left low part of screen observe the new instrument sensitivity /div on right side window.

To change the system parameters: Select MASS, SPRING or DAMPER in the MAIN MENU, select from the SLAVE MENU the parameter of element 1 or element 2. Use + to increase the parameter by one step, use - to decrease the parameter by one step select STEP in the slave menu to change the step. Use + to increase the step Use - to decrease the step. Observe values of the parameters on right side parameter window. Select MAIN and SLAVE menus with keypad arrows.

For Natural Vibration: Select NATURAL from the main menu. Select I.C. from the slave menu. Enter initial velocities and displacements, as requested. Hit ENTER after each entry. Select MAIN and SLAVE menus with keypad arrows.

To specify external forces: Select FORCED from the main menu. Select HARMONIC or PERIODIC or GENERAL from the slave menu. Hit ENTER.

If your selection is HARMONIC: You can specify up to 10 harmonic forces per mass station. For everyone enter amplitude, phase angle, frequency. Enter data as requested and hit ENTER.

If your selection is PERIODIC: You must have a text file with the values of the periodic force at n points at time steps Δt for a full period. Numbers in the same line are separated by commas. The file format is:

```
1rst line:    myfile        any identification text
2nd line:     Δt,n          time step, number of samples
3rd line:     f1,f2,......  n samples-values of the forcing function
```

Enter file name as requested.

If your selection is GENERAL f(t): You must have a prepared file, as above, for the whole record of the forcing function.

Use consistent system of units for your data. Select MAIN and SLAVE menus with keypad arrows.

To specify base motion: Select BASE EXC from the main menu. Select HARMONIC or PERIODIC or GENERAL from the slave menu. Hit ENTER.

If your selection is HARMONIC: You can give up to 10 harmonic motions of the bases. For everyone enter amplitude, phase angle, frequency. Enter data as requested and hit ENTER.

If your selection is PERIODIC: You must have a text file with the values of the periodic motion at n points at time steps Δt for a full period:

> 1rst line: myfile   any identification text (in quotes)
>
> 2nd line: $\Delta t, n$ time step, number of samples
>
> 3rd line: y1,y2,... n equidistant samples-values of the base motion

Enter file name as requested.

If your selection is GENERAL f(t): You must have a prepared file, as above, for the whole record of the base motion. Use consistent system of units for your data. Select MAIN and SLAVE menus with keypad arrows.

In the upper part of the screen are bar-graphs of the amplitudes:

> Base motion
>
> Motion of mass 1
>
> Motion of mass 2

On the left of the vibrating system, the graphs show the amplitude vs time signals. On the right of the vibrating system, the graphs show the phase portraits of the motion. On the right of the screen there is the parameter window.

EXAMPLE: Using the simulated experiment LINSYS, compute the motion of an electronic equipment with mass m = 2 kg elastically supported with k = 10 N/m, c = 5 Nsec/m in a vehicle which crashes when the deceleration of the vehicle is known from experiments to be:

```
Time (sec):         .1   .2   .3   .4   .5   .6   .7   .8   .9   1.
Acceleration (g):  0    .5   1.  1.5 2.0 1.5  1.  .5    0   0
```

The equation of motion is

$$mz'' + cz' + kz = -my''$$

where z = x - y, x the motion of the mass and y the motion of the base (Dimarogonas & Haddad 1992]. We prepare a file with the following data:

```
Test problem           ' identification
10,.1                  ' Number of records, time step
0,1,2,3,4,3,2,1,0,0  ' 10 values of -my":
```

We save it with the name C5_1.DAT

From the SIMULAB menu we select LINESYS and hit ENTER. The title page appears on the screen. We select one-mass system. Then, the menu page appears on the screen with the simulated 1 d-o-f system. With the keypad arrow keys we select menus.

a. We set mass = 2 kg, k = 10 N/m and c = 5 Nsec/m using the + , - keys.

b. We specify base excitation, non-harmonic. At the prompt we enter the file name C5_1.DAT

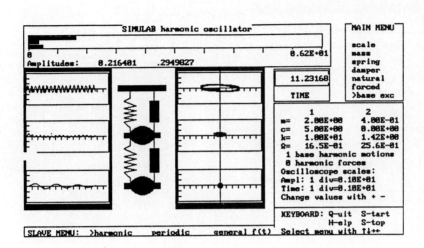

d. We hit S. The operation of the system starts. The resulting motion is shown on the screen.

Probable Errors: Absence of the COMPUME.SYS file. Division by zero or overflow, if the input is wrong. Wild graphs if scales are not selected properly.

6.3. BALANCER

Function: Simulation of a balancing machine

Limitations: 2-mass balancing.

Files Needed: BALANCER.EXE, COMPUME.SYS

Method: A turbine rotor was be modeled by way of a three-mass system m_1, m_2, m_3, connected with a massless shaft of diameter d. The two end masses represent the journals which are supported by linear bearings having stiffnesses k_1, k_2 and linear viscous damping constants c_1, c_2. The left bearing and the mid-span are the locations of the balance planes. The system is modeled with 3 dof as shown.

The vertical spring constant of the shaft itself in respect to rigid bearings is, as a simply supported beam k. If the masses are displaced by x_1, x_2, x_3, as shown, the force on the mass m_2 due to the deflection of the shaft only will be $-k[x_2-(x_1+x_3)/2]$ and on the masses m_1 and m_3 equal forces $k[x_2-(x_1+x_3)/2]/2$, due to symmetry.

Application of Newton's Law in the vertical direction, with the nomenclature on figure E8.3, yields the differential equations of motion:

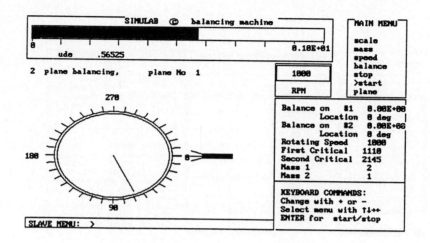

$$m_1 x^{\circ\circ}_1 + c_1 x^{\circ}_1 + (k/4 + k_1)x_1 - kx_2/2 + kx_3/4 = m_1 r_1 \omega^2 \cos(\omega t + \phi_1)$$

$$m_2 x^{\circ\circ}_2 \qquad\qquad -kx_1/2 + kx_2 \quad - kx_3/2 = m_2 r_2 \omega^2 \cos(\omega t + \phi_2)$$

$$m_3 x^{\circ\circ}_3 + c_3 x^{\circ}_3 \qquad +kx_1/4 - kx_2/2 + (k/4 + k_3)x_3 = 0$$

in matrix form

$$\underline{M}x^{\circ\circ} + \underline{C}x^{\circ} + \underline{K}x = \underline{F}_0 e^{i\omega t}$$

where

$$\underline{M} = \begin{bmatrix} m_1 & 0 & 0 \\ 0 & m_2 & 0 \\ 0 & 0 & m_3 \end{bmatrix}, \quad \underline{C} = \begin{bmatrix} c_1 & 0 & 0 \\ 0 & 0 & 0 \\ 0 & 0 & c_3 \end{bmatrix}, \quad \underline{K} = \begin{bmatrix} k/4 + k_1 & -k/2 & k/4 \\ -k/2 & k & -k/2 \\ k/4 & -k/2 & k/4 + k_3 \end{bmatrix},$$

$$\underline{x} = \{x_1 \ x_2 \ x_3\}, \quad \underline{F}_0 = \{m_1 r_1 \omega^2 (\cos\phi_1 + i\sin\phi_1) \quad m_2 r_2 \omega^2 (\cos\phi_2 + i\sin\phi_2) \quad 0\}$$

The solution is given by equations 8.10:

$$[-\omega^2 \underline{M} + i\omega \underline{C} + \underline{K}]\underline{X} = \underline{F}_0$$

This is a system of complex linear algebraic equations:

$$\begin{bmatrix} -m_1\omega^2 + i\omega c_1 + k/4 + k_1 & -k/2 & k/4 \\ -k/2 & -m_2\omega^2 + k & -k/2 \\ k/4 & -k/2 & -m_3\omega^2 + i\omega c_3 + k/4 + k_3 \end{bmatrix} \begin{bmatrix} X_1 \\ X_2 \\ X_3 \end{bmatrix} = \begin{bmatrix} F_1 \\ F_2 \\ 0 \end{bmatrix}$$

The solution is obtained with Cramer's Rule:

$$
X_1 = \begin{vmatrix} -m_1\omega^2+i\omega c_1+k/4+k_1 & F_1 & k/4 \\ -k/2 & F_2 & -k/2 \\ k/4 & 0 & -m_3\omega^2+i\omega c_3+k/4+k_3 \end{vmatrix} / D
$$

$$
X_2 = \begin{vmatrix} F_1 & -k/2 & k/4 \\ F_2 & -m_2\omega^2+k & -k/2 \\ 0 & -k/2 & -m_3\omega^2+i\omega c_3+k/4+k_3 \end{vmatrix} / D
$$

where

$$
D = \begin{vmatrix} -m_1\omega^2+i\omega c_1+k/4+k_1 & -k/2 & k/4 \\ -k/2 & -m_2\omega^2+k & -k/2 \\ k/4 & -k/2 & -m_3\omega^2+i\omega c_3+k/4+k_3 \end{vmatrix}
$$

Use: On the title page, the program prompts for one or two balancing planes, and for input or random unbalance. If the user selection is "input", the program prompts the user to enter unbalance by magnitude and orientation. If the selection is "random" then the program uses a random number generator to assign unbalance. The random values selected will be reported at the end of the session. Finally, the program prompts for sound (Y/N). If the answer is Y then during operation the program generates a sound with pitch proportional to the speed of rotation.

Then, the main screen appears. On the right there is the main menu and on the screen bottom is the slave menu, submenu of the main menu. Selection is done with the keypad arrows.

The system on the screen has two states: rotation and no-rotation. To start rotation, select <start> in the main menu and hit ENTER. To stop rotation, select <stop> in the main menu and <operation. in the slave menu and hit ENTER. In the state of rotation, the stroboscope flashes, the instruments show readings and, if sound was selected, sound is generated with frequency proportional to the rotational frequency.

To change the system parameters: With non-rotation state, select <mass>, <speed>, or <balance>. in the main menu, select from the slave menu the parameter you want to change. Use + to increase the parameter by one step, Use - to decrease the parameter by one step select STEP in the slave menu to change the step. Use + to increase the step Use - to decrease the step. Observe values of the parameters on right side parameter widow. Selection <plane> affects the plane where instrument readings are shown and balance plane where the balance mass is placed. You can also change plane by hitting ESC.

Select <scale> to change scale if the readings are small or out of scale.

In the upper part of the screen are bar-graphs of the amplitudes. On the right are the problem parameters. On the polar graphs are the location of the unbalance, in magnitude and direction (diameter and orientation of the red circle, respectively) and the phase angle.

Example:

We select 2-plane balancing, random input, no sound. The main screen appears. We select <start> and hit ENTER. Operation starts. We hit ESC to view the measurements on both planes. We record the measurements.

We select <stop>, <<operation>> and hit enter. Operation stops. We select <plane> <plane 1> and hit enter (or hit the ESC key). We select <balance>, <<change>> and hit the + key until the reading for unbalance on plane #1 is 1. We select <start> and hit ENTER. We change planes (or use ESC) to obtain the response in both planes.

We now select <stop> <<rotation>>, hit ENTER and return to plan 1. We select <balance> <<change>> and hit the - key until the added unbalance is 0 in plane 1. We change to plane 2 and increase the unbalance in this plane to 1. We start and measure the response in both planes, as above. We record the measurements.

We now select <stop>, <<execution>> and hit ENTER. Execution is terminated and the program prompts for a file name to write the results. BALANCER.RES is the default. A complete record of all starts is presented on the screen and written in the selected file. For each start, the first line corresponds to plane 2 and the second line to plane 2:

```
BALANCING REPORT
Initial unbalance:
Plane 1: unbalance =   .3147471  @ angle  50.56438  deg.
Plane 2: unbalance =   .5170268  @ angle  4.939234  deg.

start amplitude    phase angle  balance shot   angle
  1   .3206789        -25          0             0
      .2051712         68          0             0
  2   .9403667         33          1             0
      .5005949       -126          0             0
  3   .9232122       -107          0             0
      .8985517         47          1             0
```

Probable Errors: Absence of the COMPUME.SYS file. Division by zero or overflow, if the input is wrong. Wild graphs if scales are not selected properly.

6.4. **POINCARE** map and dynamics of nonlinear systems.

Function: Phase Portrait and Poincare Map of a non-linear Oscillator.

Files Needed: POINCARE.OBJ, POINCFUN.OBJ, COMPUME.SYS

Method: Suppose that we take a photograph of the phase portrait of a nonlinear system at time intervals equal to the period of oscillation of the pendulum. If we do the same with the same frequency but at different locations, say at angles $\pi/6$ apart, the trace of the response on the phase portraits will not be the same. If the response is chaotic, at every frame there will be no single point but several because the three-dimensional response curve is different at each cycle. The plot on each of the frames is called a *Poincare' Map*.

The nonlinear differential equations are written in normal form:

$$\underline{x}^\circ = \underline{f}(\underline{x}, t)$$

These equations are solved numerically. The Euler, Runge-Kutta and Runge-Kutta-Merson methods of integration are used.

Use: You should describe the nonlinear system in the source code POINCFUN.HLL, where HLL is the MICROSOFT high level langauage of your choise, BAS, FOR, C for QuickBasic, Fortran and C respectively. Define in the subroutine DERIV, the first derivatives evaluation. To this end, the program uses wx(1), wx(2),... as variable names and expects the derivatives to be computed as wD1, wD2, ..., respectively. Use any parameter names, but preferably start all your variables with w to avoid conflict with program parameters in COMMON.

Compile the program POINCFUN.HLL to the object code POINCFUN.OBJ and link it with POINCARE.OBJ:

```
C:\LINK POINCARE.OBJ+POINCFUN.OBJ,  C:\MELAB\POINCARE.EXE
```

Run the resulting POINCARE.EXE from the COMPUME menu. The user is given the choice of a) Phase Portrait or Poincare' Map, b) Integration method: Modified Euler, Runge-Kutta, Runge-Kutta-Merson Methods.

Poincare plots the response in the plane ($\phi-\Omega\tau$, $\phi^\circ-\Omega$) to be able to plot torsional vibration because the angle of rotation increases continuously. For systems without this feature, the variable **om** is simply put equal to zero and then the program plots in the plane (x, x°). For multi-degree of freedom systems, only the first two coordinates ($x_1=x_2^\circ$, x_2) and ($x_3=x_4^\circ$, x_4) are plotted in different colors. Up-to 5 degrees of freedom are allowed.

EXAMPLE: Consider, for example, the turning gear mechanism of turbine rotors (Dimarogonas & Haddad 1992). During heating or cooling, the machine is set in slow rotation to avoid creep deformations by a motor and a gear couple. The gear teeth have a gap b. The equations of motion of the system are, assuming that ϕ is the angle of rotation, the angular velocity of the motor is constant $R\Omega$ and of the turbine Ω, where R is the gear ratio,

$$J\phi^{\circ\circ} + F(\phi) = sgn(\phi^\circ - \Omega)\mu mg$$

```
where F(φ) = k(φ-Ωt) if (φ-Ωt) > 0
           = 0 if -b < (φ-Ωt) < 0
           = k(φ-Ωt-b) if (φ-Ωt) < -b
       μ = e^1-(φ°-Ω)/Ω
```

In normal form

$$x^\circ_1 = -[F(\phi) + sgn(\phi^\circ-\Omega)\mu mg]/J$$
$$x^\circ_2 = x_1$$

The program POINCARE will be used with J=10000 kgm^2, k=10000 Nm/rad, Ω=0.1, μ=.1e$^{1-(\phi^\circ-\Omega)/\Omega}$ rad/sec, b=0.01 rad, mg=10000 N. Chaotic motion is observed, figure (a). Increasing the rotation frequency to 1 rad/sec results in a limit cycle, figure (b).

The program POINCFUN.BAS, which is written by the user, follows:

```
REM   The user should write ONLY what is between dotted lines
REM   in subroutine DERIV
END
```

```
SUB DERIV (xe(), de(), t, dt, om) STATIC
REM *****************************DERIVATIVES EVALUATION
'
' Here starts the user-written program
' ......................Parameter definition:
      x1 = xe(1): x2 = xe(2)
'......................................... system data
     Jiner = 10000!: stiff = 10000!: b = .01: Force = 0: om = .1
     dt = (6.28 / om) / 400
     ex = 1 - (x1 - om) / om:
     mumg = 10000! * EXP(ex)
     IF mumg > 10000 THEN mumg = 10000
     kx = 0
     IF x2 - om * t > 0 THEN kx = stiff * (x2 - om * t)
     IF x2 - om * t < -b THEN kx = stiff * (x2 - om * t - b)
'...................................derivatives evaluation
     d1 = -(kx + SGN(x1 - om) * mumg) / Jiner
     d1 = d1 + (-c * x1 + Force * COS(om * t)) / Jiner
     d2 = x1
     de(1) = d1: de(2) = d2: de(3) = d3: de(4) = d4 '...derivatives vector
' Here stops the user-written program
' -------------------------------------------------------------------

END SUB
```

Probable Errors: Absence of the COMPUME.SYS file. No trace on the screen, means that the xsc parameter was not properly selected in the PARAMETERS subroutine. You can use also the increase or decrease scale commands (< or >) until the trace is in the desired scale. Or, you did not specify forcing function or initial conditions in your system of equations.

Wild variations of the trace probably mean very high integration step and numerical instability. Try using the R-K-R integration method which has adjustable step.

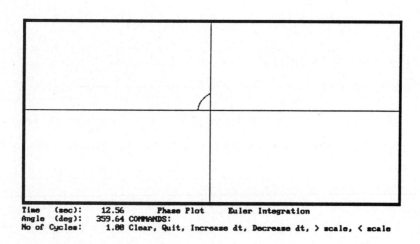

```
Time    (sec):   12.56     Phase Plot    Euler Integration
Angle  (deg):   359.64 COMMANDS:
No of Cycles:    1.88 Clear, Quit, Increase dt, Decrease dt, > scale, < scale
```

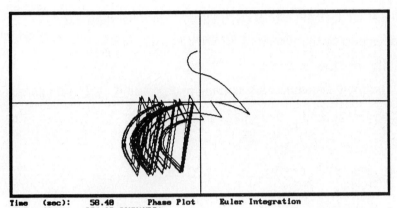

Time   (sec):   58.48   Phase Plot   Euler Integration
Angle (deg):   334.63 COMMANDS:
No of Cycles: 8.88 Clear, Quit, Increase dt, Decrease dt, > scale, < scale

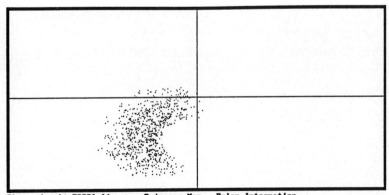

Time   (sec): 58886.61   Poincare Map   Euler Integration
Angle (deg):   318.89 COMMANDS:
No of Cycles: 889.88 Clear, Quit, Increase dt, Decrease dt, > scale, < scale

6.5. MULTI-degree-of-freedom LINear system analysis.

Function: Dynamics of a multi-degree-of-freedom linear system. Eigenvalue problem, response to harmonic excitation, transient response with the modal analysis method.

Limitations: 10 degrees of freedom.

Files Needed: MULTILIN.EXE, COMPUME.SYS, SYS.DAT or other data file.

Method: The system is described by the system of differential equations

$$\underline{M}x" + \underline{C}x' + \underline{K}x = \underline{f}(t)$$

Assuming that the modal matrix Φ decouples the damping matrix, and $\Phi^T\underline{C}\Phi$ is diagonal,

$$q_i(t) = e^{-\zeta_i\omega_i t}\{q_i(0)\cos\omega_{di}t + [2\zeta_i\omega_i q_i(0) +$$

$$+ q'_i(0)/\omega_{di}]\sin\omega_{di}t\} + \int_0^t F_i(t*)g_i(t - t*)\Delta t*$$

$$\text{where}\quad \omega_{di} = \omega_i^2(1-\zeta_i^2)^{1/2},$$

$$g_i(t) = (1/\omega_{di})e^{-\zeta_i\omega_i t}\sin\omega_{di}t,$$

$$F_i(t) = \phi_i^T\underline{f}(t)$$

$$\underline{q}(0) = \underline{\phi}^T\underline{M}x(0)$$

$$\underline{q}'(0) = \underline{\phi}^T\underline{M}x'(0)$$

$$\underline{x} = \underline{\Phi}q$$

and ζ_i is the modal damping factor [Dimarogonas & Haddad 1992].

Use: The module is selected from the root menu of MENUMEC. The menu selections of MULTILIN are:

Quit: Quits the program execution and returns to COMPUME.

Read: Reads the input data file. SYS.DAT is a default file for 3 degrees-of-freedom:

```
Linear System Data File sys.dat
Number of degrees of freedom/ No of Angular Velocities/ angular velocities
 3 , 1, 100
Mass Matrix
 1 , 0 , 0
 0 , 1 , 0
 0 , 0 , 1
Damping Matrix
 0 , 0 , 0
 0 , 0 , 0
 0 , 0 , 0
Stiffness Matrix
 4 , 2 , 2
 2 , 5 , 1
 2 , 1 , 6
Force vector
0,0,1
Phase angle vector
0,0,0
```

```
Initial Displacements
0 , 0 , 0
Initial velocities
0 , 0 , 0
Number of frequencies for modal analysis
3
```

Input: Editor to edit the input file or to make a new data file.

Save: Saves the data file.

HarmExc: Solves the harmonic excitation problem:

$$\underline{x} = [\underline{K} - \omega^2 \underline{M}]^{-1} \underline{F}$$

Eigenvalue: Solves the eigenvalue problem

$$[\underline{K} - \omega^2 \underline{M}] = \underline{0}$$

with the Cholesky/Jacobi and the Stodola/power iteration methods.

Modal: Finds the response of the system with given initial conditions with the modal analysis method.

Dir: Prints a directory of available data files.

EXAMPLE: The response of the above system in the SYS.DAT file, written in the results file SYS.RES, is as follows:

```
System Data: SYS.DAT
 Mass Matrix
    0.10E+01    0.00E+00    0.00E+00
    0.00E+00    0.10E+01    0.00E+00
    0.00E+00    0.00E+00    0.10E+01
 Damping  Matrix
    0.00E+00    0.00E+00    0.00E+00
    0.00E+00    0.00E+00    0.00E+00
    0.00E+00    0.00E+00    0.00E+00
 Stifness  Matrix
    0.40E+01    0.20E+01    0.20E+01
    0.20E+01    0.50E+01    0.10E+01
    0.20E+01    0.10E+01    0.60E+01
 Number of excitation frequencies:    1
 Excitation frequencies: 100    rad/sec.
 Excitation:
 No of Force Vectors:    1
 Force vector 1
 No        Amplitude        Phase Angle (deg)
  1        0.00E+00         0.00E+00
  2        0.00E+00         0.00E+00
  3        1.00E+00         0.00E+00
 Solution:
 No        Xreal            Ximaginary
  1        -1.00E-01        0.00E+00
  2        0.00E+00         0.00E+00
  3        2.00E-01         0.00E+00
```

Probable Errors: Absence of the COMPUME.SYS file. Subscript out of range, if the problem is larger than allowed (10 dof). Division by zero or overflow, if an input matrix is singular.

7

CHAPTER SEVEN
ROTOR DYNAMICS

7.1. ROtor DYNAmic analysis

Function: Dynamic analysis of rotors.

Limitations: 10 elements.

Files needed: RODYNA.EXE, COMPUME.SYS, PARAM.TMP, ROTOR1.EXE, modules ROTOR2.EXE
ROTOR8.EXE, depending on the options required.

Method: The rotor is modeled by the finite element method in the form

$$\underline{M}\mathbf{x}'' + \underline{C}\mathbf{x}' + \underline{K}\mathbf{x} = \underline{F}e^{i\omega t}$$

where **x** is the vector of lateral displacements and slopes, for lateral vibration analysis, or the angles of twist, f
torsional vibration analysis [Dimarogonas & Haddad 1992, chapter 11].

```
 Quit  Files  Plot   Critic Linear Bearg  Nonlin Specan Model  Develo   RODYNA
       Directory
       Load File
       Save File
```

File manipulations   Hit F1 for help

Use: The module is selected from the COMPUME menu. A menu bar appears on top of the screen with the program options. You highlight a program option with the horizontal keypad direction keys or by typing the first letter of the option, ie **F** for Files. A submenu rectangular window is shown below the menu bar option. Using the keypad vertical direction keys you highlight a submenu window option. Hitting the ENTER key runs the program. When execution is terminated, control returns to the menu bar.

Now you are ready to select the RODYNA options from the menu. Every time a RODYNA module is terminated, control returns to the RODYNA main menu.

In the following, the pull-down menu options are described:

7.1a.	Quit

Function: Quit the RODYNA execution

Method: Exits to operating environment (DOS, UNIX, High Level Language, etc.). Closes all open files. Resets temporary files.

Limitations: None

Use: With <Quit> you will:

a. Exit RODYNA and return to the operating environment from which RODYNA was loaded (DOS, UNIX, High Level Language, etc.).

b. All changes in the rotor model will be lost if you did not save them with the <save> command.

c. Some information resides in temporary files, such as the critical speeds computed last are in the file CRITICAL.TMP, for example. You could retrieve such information with any text editor (ASCII files).

7.1b	Dir

Function: File Directory

Method: Scans current directory with rotor files (with .ROT extension).

Limitations: At least one .ROT file must be in the current directory.

Use: With <Dir> you will have a list of files with extension .ROT with rotor models already stored in the current drive.

If your file does not show in this directory,

a. You might have stored it in another drive or directory. Use <Load> with the wild character * to search other drives or sub-directories, ie,

    select <Load>
    Hit   <Enter>
    Answer <a:\oldones*.xyz> to search drive A, directory \oldones
                            for files with extension .xyz

If you want to make the model of a new rotor, use <Model>

7.1c.	Load

Function: Load File

Method: Loading a data file from current drive.

Limitations: None.

Use: With <Load> you can:

1. Load a file which was stored previously:

enter full name (with extension) after the prompt ?

If you already have a rotor in memory, RODYNA will ask you to decide:

If you want to add the new rotor on the right of the existing one (A)

If you want the old rotor to be erased and the new one loaded (N)

If you want to cancel the <Load> command (C)

2. Have a list of files with any name or extension with rotor models already stored in the current drive, using wild characters.

To do this, for example, answer   A:\OLDONES*.XYZ   to search drive A, directory \OLDONES for files with extension .XYZ

If you want to make the model of a new rotor, use <Model>. A data file can also be prepared in a word processor. The file format is:

```
Problem identification
PROPERTIES, No of elements, Young's modulus, density, gravity constant
ELEMENT, 1, length, diameter
(as many lines as elements, consecutively numbered starting from the left
end)
..............................................
NODE, 1, unbalance (mass x radius), disk mass, disk inertia,
                bearing index,0,0,0,0, bearing stiffness, bearing
damping,0,0
(as many lines as nodes, consequtively numbered starting from the left end)
..............................................
FILEND
```

Bearing index is 1 if there is a bearing, 0 otherwise (underlined entries omitted).

A typical such file, HP.ROT, follows:

```
High Pressure Turbine Rotor
PROPERTIES, 8 , 2.1E+11 , 7800 , 9.81
ELEMENT, 1 , .8 , .3
ELEMENT, 2 , 1.5 , .35
ELEMENT, 3 , 1 , .4
ELEMENT, 4 , 1 , .45
ELEMENT, 5 , 1 , .5
ELEMENT, 6 , 1 , .55
ELEMENT, 7 , 1 , .5
ELEMENT, 8 , .8 , .5
NODE, 1 , 0 , 50 , .5 , 0 , 0 , 0 , 0
NODE, 2 , 0 , 50 , .2 , 1 , 0 , 0 , 0 , 0 , 3.879232E+08 , 1724710 , 0 , 0
NODE, 3 , 0 , 500 , 50 , 0 , 0 , 0 , 0
NODE, 4 , 0 , 700 , 80 , 0 , 0 , 0 , 0
```

```
NODE, 5 , .0001 , 1100 , 200 , 0 , 0 , 0 , 0 , 0
NODE, 6 , 0 , 1300 , 250 , 0 , 0 , 0 , 0
NODE, 7 , 0 , 1500 , 300 , 0 , 0 , 0 , 0
NODE, 8 , 0 , 50 , .5 , 1 , 0 , 0 , 0 , 0 , 3.879232E+08 , 1724710 , 0 , 0
NODE, 9 , 0 , 600 , 40 , 0 , 0 , 0 , 0
FILEND
```

7.1d.   Save

Function: Save File

Method: Saving a data file to the current or a specified dive.

Limitations: None.

Use: With <Save> you can:

1. Save a file which has been created new or modified with <Model>: enter full name (with extension) after the prompt ? and hit ENTER...

2. Just hit ENTER to exit <Save> if you changed your mind.

3. Have a list of files with any name or extension with rotor models already stored in the current drive, using wild characters.

To do this, answer   A:\OLDONES*.XYZ (ENTER) to search drive A, directory \OLDONES for files with extension .XYZ

This will help you avoid storing the current model to an existing file name because then the old file will be lost.

7.1e.   ClearScr

Function: Clear Screen

Method: Erases the screen below the bar menu.

Limitations: None.

Use: With <Clear Screen> you can erase the working part of the screen.

1. The model in memory is not erased.

2. You can restore the rotor plot by using <Plot>.

3. The data on the screen will be lost. To recover them, you might have to run the particular option again.

7.1f.   Plot

Function: Plot Rotor.

Method: Plotting the rotor, supports, disks, unbalance.

Limitations: Very complicated geometries might not show well for low screen resolutions.

Use: 1. Use <Plot> to plot the shape and the loading conditions on the rotor.

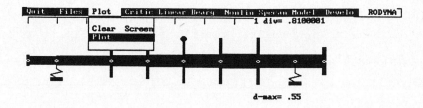

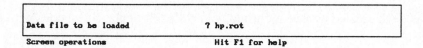

2. Dimensions are on scale. Vertical and horizontal scale is not exactly the same, since it depends on the screen. The scales are shown on the screen.

3. Bearings are shown as springs. Dampers are not shown on the figure.

4. Unbalances are shown as red dots on the disks on the nodes.

5. Node disk diameters have been computed from the mass and the moment of inertia, assuming constant thickness. For this reason, some look strange.

6. Nodes are indicated with dark circles along the rotor centerline.

7. Plotting the rotor erases the screen.

8. The program plots the rotor in some cases (ie vibration nodes) by itself.

9. To view other conditions (misalignment, cracks, rubbing, etc, use <Make>).

10. After the use of <Model>, always plot the rotor to make sure that geometric data are correct.

EXAMPLE: A typical high-pressure turbine rotor was loaded from file HP.ROT and plotted as shown with the <Plot> command.

7.1g.	Torsional

Function: Torsional Natural Frequencies of Shafts.

Method: Finite Element modelling, using cylindrical elements. The natural frequencies and vibration modes are computed with the Jacobi or the Power Iteration method, they are ordered and the natural modes are normalized.

Limitations: The number of nodes is limited in this version to 11.

Use:

1. <Torsional> computes the torsional natural frequencies and torsional vibration modes for the rotor in memory. Before you run <Torsional>:

2. Use <Plot> to plot the shape and the loading conditions on the rotor in memory. If <Plot> shows nothing, it means that there is no rotor in memory. Go back to the <files> menu and <load> a rotor or go to the <Model> menu and enter the data for a new rotor.

3. The program assumes prismatic elements between nodes. The frequencies and modes are solutions of the eigenvalue problem:

$$[M]\{x''\} + [K]\{x\} = 0$$

4. Two methods of solution are available:

a. Jacobi Rotation method. It computes concurrently ALL natural frequencies and modes. Excellent for small, slow for large systems.

b. Power Iteration method. Obtains a number of frequencies, specified by the user, by iteration. Better for large systems.

5. The natural frequencies and vibration nodes are stored, for the last analysis performed, on the file CRITICAL.TMP on the current drive.

6. Always plot the modes of interest. Abrupt shapes might indicate non-convergence. This is normal for very high modes and appears sometimes at lower modes for certain ill-conditioned problems. It might indicate wrong data.

7. The first mode is a parallel line. Indicates only rigid body motion. The first frequency should be zero. Numerically, it will be a very small number, much smaller than the next one.

8. The frequencies are given in ascending order. If it is not so, it indicates non-convergence or another problem. For some combinations of analyses, subsequent computation of the critical speeds might present problems, due to the use of matrices for other purposes to conserve space.

9. The program computes torsional natural frequencies which, for torsional vibration only, are the same with the torsional critical speeds.

EXAMPLE: The torsional natural frequencies of the above rotor of file HP.ROT are to be computed. <Torsional> is selected and then <J> for Jacobi method and the shear modulus $G = 1.05 \times 10^{11}$ N/m^2 is entered after the appropriate prompt. The program computes the natural frequencies and animates the natural modes. The results are saved in the temporary file CRITICAL.TMP which should be renamed if the user wants to keep it for later reference.

```
Torsional Natural Frequencies of last rotor
 -.76E+00, 0.11E+04, 0.18E+04, 0.27E+04, 0.31E+04, 0.37E+04, 0.37E+04, 0.91E+04, 0.12E+05 rad/sec
0.29E-01, -.88E-01, -.11E+00, -.13E+00, 0.19E+00, 0.23E+00, 0.75E-01, 0.59E+00, 0.14E-06 Hz
Modes
 0.29E-01, -.85E-01, -.98E-01, -.11E+00, 0.14E+00, 0.15E+00, 0.48E-01, -.26E+00, -.13E-06,
 0.29E-01, -.71E-01, -.55E-01, -.14E-01, -.18E-01, -.77E-01, -.26E-01, 0.18E-01, 0.17E-06,

 0.29E-01, -.44E-01, 0.94E-02, 0.49E-01, -.48E-01, 0.47E-01, 0.20E-01, -.13E-02, -.19E-05,

 0.29E-01, -.14E-01, 0.42E-01, 0.77E-02, 0.36E-01, -.11E-01, -.11E-01, 0.65E-04, 0.21E-04,
 0.29E-01, 0.11E-01, 0.12E-01, -.39E-01, -.24E-01, -.91E-03, 0.16E-01, -.38E-05, -.38E-03,
 0.29E-01, 0.25E-01, -.21E-01, 0.13E-01, 0.27E-02, 0.98E-02, -.27E-01, 0.27E-06, 0.58E-02,
 0.29E-01, 0.28E-01, -.31E-01, 0.41E-01, 0.29E-01, -.37E-01, 0.90E-01, 0.58E-07, -.16E+00,
 0.29E-01, 0.29E-01, -.34E-01, 0.49E-01, 0.37E-01, -.52E-01, 0.13E+00, -.31E-06, 0.25E+00,
```

The modes are shown in the following figures:

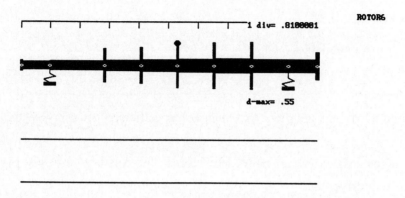

ROTOR6

1 div= .8100001

d-max= .55

y-max= 2.938035E-02
Critical Speed 1 : -.7600626 rad/s   Hit ENTER for next mode, E to exit

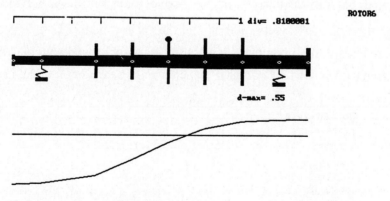

ROTOR6

1 div= .8100001

d-max= .55

y-max= 2.871434E-02
Critical Speed 2 : 1898.758 rad/s   Hit ENTER for next mode, E to exit

7.1h.	Lateral

Function: Lateral Natural Frequencies of Rotors.

Method: Finite Element modelling, using Cylindrical Elements. The Natural frequencies and vibration modes are computed with the Jacobi or the Power Iteration method, they are ordered and the natural modes are normalized.

Limitations: The number of nodes limited to 11 in this version.

Use:

1. <Lateral> computes the lateral natural frequencies and lateral vibration modes for the rotor in memory. Before you run <Lateral>:

2. Use <Plot> to plot the shape and the loading conditions on the rotor in memory. If <Plot> shows nothing, it means that there is no rotor in memory. Go back to the <files> menu and <load> a rotor or go to the <Model> menu and enter the data for a new rotor.

3. The program assumes prismatic elements between nodes. You can use the following 2 mass models in response to the appropriate prompt:

    a. Consistent mass: Mass of the element itself, uniformly distributed. Modelled with the Finite Element Method. Suggested.

    b. Lumped: Mass of the element itself, lumped at the element ends. Stiffness modelled with the Finite Element Method. The frequencies and modes are solutions of the eigenvalue problem:

$$[M]\{x''\} + [K]\{x\} = 0$$

4. Two methods of solution are available:

    a. Jacobi Rotation method. It computes concurrently ALL natural frequencies and modes. Excellent for small, slow for large systems.

    b. Power Iteration method. Obtains a number of frequencies, specified by the user, by iteration. Better for large systems.

5. The natural frequencies and vibration nodes are stored, for the last analysis performed, on the file CRITICAL.TMP on the current drive.

6. Always plot the modes of interest. Abrupt shapes might indicate non-convergence. This is normal for very high modes and appears sometimes at lower modes for certain ill-conditioned problems. It might indicate wrong data, ie missing bearings.

7. The frequencies are given in ascending order. If it is not so, it indicates non-convergence or another problem. For some combinations of analyses, subsequent computation of the critical speeds might present problems, due to the use of matrices for other purposes to conserve space.

8. The program computes natural frequencies at a specific speed, because the bearing properties depend on speed. Wherever in the program the words 'critical speeds' appear, they mean 'natural frequency at a given speed'.

True critical speeds can be found by iteration: Compute first the natural frequency at running speed. To determine the critical speed which corresponds to a certain natural frequency, compute the bearing constants at this frequency and repeat the computation, until assumed and computed natural frequency is the same.

EXAMPLE: For the rotor of file HP.ROT find the lateral natural frequencies.

<critical> is selected and then <D> for distributed mass and <J> for the Jacobi method. The results are stored in file CRITICAL.TMP. The natural frequencies are:

```
Critical speeds of last rotor
  92.17628 , 271.7198 , 407.9361 , 555.2486 , 789.4803 , 1279.743 , 1849.425
, 2414.265 , 2981.191 , 3880.308 , 4483.984 , 5052.172 , 5784.295 , 6028.896
, 6847.417 , 8116.429 , 16073.32 , 17246.48 ,rad/sec
```

The lowest modes are shown here:

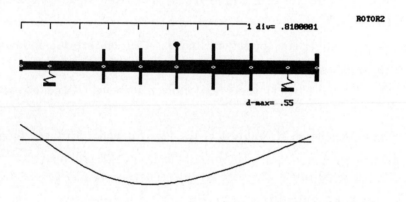

Critical Speed 1 : 92.17628 rad/s   Hit ENTER for next mode, E to exit

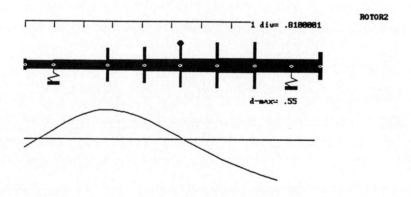

Critical Speed 2 : 271.7198 rad/s   Hit ENTER for next mode, E to exit

7.1i.   Long

Function: Longitudinal Natural Frequencies of shafts.

Not included in this version.

7.1j.   Linear

Function: Dynamic Analysis of Linear rotors with the Finite Element Method.

Method: Finite Element modelling, using cylindrical elements. A banded system of complex algebraic equations is solved with Gauss Elimination.

Limitations: 10 elements.

Use:

1. Linear rotor and linearized bearing coefficients are used for the computation of the rotor response to the unbalance specified in the rotor model definition.

2. The program asks for the angular velocity of rotation.

3. The Finite Element Method and the consistent mass model are used.

4. Plot rotor first to make sure that you have a rotor in memory. If <Plot> shows nothing, go to <Files> and <Load> a rotor or use <Model> to define a new one.

5. Observe the boundary conditions (zero moments at the ends). If not so or the vibrating rotor shape is diverging from left to right, it shows numerical problems or the data are wrong. Use <Model> to view the data. If the data are correct and the problem persists, it is numerical. Try using different units.

6. Maximum rotor deflection is given on the animated vibration mode plot.

EXAMPLE: Find the response of the rotor in the file HP.ROT to the unbalance given in the file.

The **Model** selection of the menu invokes the spreadsheet input:

```
 Quit   Files  Plot   Critic Linear Bearg  Nonlin Specan Model  Develo   ROTOR1
High Pressure Turbine Rotor
              1         2         3         4         5         6         7
NODE DATA:
Local Mass    50        50        500       700       1100      1300      1500
Moment of Inerti>.5      .2        50        90        200       250       300
Unbalance     0         0         0         0         .0001     0         0
Bearing Stiff X 0       0.39E+09  0         0         0         0         0
Bearing Damp C 0        0.17E+07  0         0         0         0         0
Reserved      0         0         0         0         0         0         0
Coupling y    0         0         0         0         0         0         0
Coupling theta 0        0         0         0         0         0         0
ELEMENT DATA
Length        .8        1.5       1         1         1         1         1
Diameter      .3        .35       .4        .45       .5        .55       .5
Reserved,     0         0         0         0         0         0         0
Reserved"     0         0         0         0         0         0         0


Enter Material Data:             Young Modulus  2.1E+11            ?
                                 Density        7800               ?
                                 Gravity const  9.81               ?
                                        Are data correct (Y/N)  ? █

Moment of inertia concentrated at the node (Disk)           Hit X to exit
```

The response is shown in the RODYNA output:

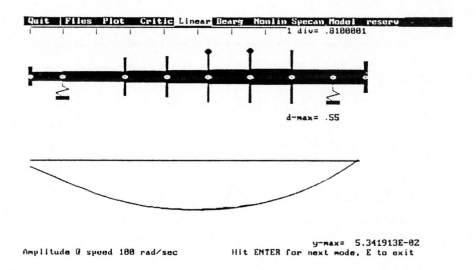

7.1k.	Balancing

Function: Dynamic Balancing of rotors. Not included in this version

---

7.1l.	Slider

Function: Analysis of slider bearings.

Method: Solution of the Reynolds Equation with the Finite Element method. For given operating conditions, the equilibrium position is found and then the linear spring and damping coefficients and nonlinear coefficients up to the third order.

Limitations: Program AUTOMESH must be used to create the triangular mesh. A file MESH*.DAT. must exist in the current directory before you use Slider.

Up to 100 nodes, 100 elements. Limits can be extended by appropriate changes in the program.

UTILIZATION:

To use the program (User types what below is underlined): Select <Slider> and hit ENTER.

The first window is for program identification. Hit ENTER.

Next is then the main menu window.

You are now in the main menu. You select with the keypad direction keys. The > sign indicates the selection and at the bottom line there is an explanation of the command. ENTER invokes the respective command.

Stop: Quits the program execution and returns to menu.

Load file: Loads a mesh file from disk. The mesh was generated with the AUTOMESH program.

Plot Mesh: Plots the mesh currently in memory. The mesh can be anywhere in the screen.

Boundary: To specify boundary conditions. Its menu:

N finds the nearest node to the cross cursor. This cursor moves with the keypad direction keys.

The following two commands are related to the motion of the cursor cross:

F moves the cursor in larger steps (multiplication by 10)

S moves the cursor in smaller steps (division by 10)

E Exits to the main menu.

P specifies zero pressure at current node. Move cursor near the desired node and hit N. The cursor locks on the node. Hit P. A circle is placed around the node indicating zero pressure. For zero pressure on the boundary, answer Y to the question "Set Boundary to Po?" . Then the program asks for the number of nodes (partitions) in the peripheral-x and axial-y direction. Enter them. The nodes of the boundary are circled.

M when you specified all pressures on one arc, hit M to display the next arc to specify boundary conditions.

Analyze: The program performs the analysis and prints pressures at the nodes and bearing properties.

Post Proc: Plots the pressure distribution in a color code.

Make Model: To view and/or change default data. To keep default data just hit ENTER. To change data, enter new value and hit ENTER.

EXAMPLE: Find the eccentricity ratio, attitude angle and the stiffness and damping coefficients of a 150° journal bearing with diameter d = 300 mm, length (width) L = 200 mm, radial clearance R/1000, viscosity η = 0.005 Pas, vertical load 22,000 N, one symmetric arc.

The input is prepared with <Make Model> and the input file BEARING1.DAT follows:

```
"Radius-omega-clearance-viscosity-eccentr-angle-load-noarcs"
.15,377,.00015,.005,4.713871E-05,56.24197,22000,1
"arc- angle1-angle2-width-preload"
1,-75,75,.2,0
```

The BEARING1.DAT file was used to write the results. The results file follows:

```
"Radius-omega-clearance-viscosity-eccentr-angle-load-noarcs"
.15,377,.00015,.005,4.713871E-05,56.24197,22000,1
"arc- angle1-angle2-width-preload"
1,-75,75,.2,0
"Sommerfeld number"," eccentricity ratio","length/D","  angle"
.818616,.314258,.6666666,56.24197
" kxx"," kxy"," kyx"," kyy"
3.879232E+08,5.84095E+08,5.106766E+07,8.497302E+07
" cxx"," cxy"," cyx"," cyy"
1724710,-126187.6,353487.9,123154.2
" SXxx, SXyy, SYxx, SYyy, SXxy, SYxy "
2.35574E+12,5.285694E+12,1.060735E+12,2.382544E+12,1.424733E+13,6.44432E+12
"SXyyy, SYyyy, SXxxx, SYxxx, SXxxy, SYxxy, SXxyy, SYxyy"
2.912916E+17,1.583138E+17,8.605323E+16,4.659823E+16,1.296481E+17,7.03883E+1
6,1.933287E+17,1.048106E+17
```

The screens produced during the execution follow:

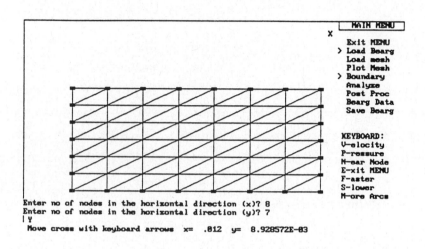

Enter no of nodes in the horizontal direction (x)? 8
Enter no of nodes in the horizontal direction (y)? 7

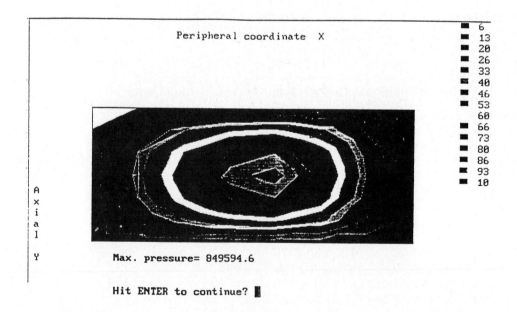

Peripheral coordinate   X

Max. pressure= 849594.6

Hit ENTER to continue?

PROBABLE ERRORS: Execution breaks: Very large mesh for the screen. Hardware incompatibility, wrong graphics adaptor. Out of memory. Absence of the COMPUME.SYS file. Execution breaks at post processing: Missing EGA card. Division by zero: data missing, you did not load mesh or boundary conditions.

7.1m.	Pad

Function: Analysis of Pad Bearings

Not included in this version of RODYNA.

7.1n.	Nonlinear

Function: Dynamic Response of Nonlinear rotor-bearing systems.

Not included in this version

7.1o.	Crack

Function: Cracked rotor elements.

Not included in this version

7.1p.	Rubbing

Function: Analysis of rotor rubbing (The Newkirk Effect)

Not included in this version.

7.1q.	**ANALYZER**

Function: Fast Fourier Transform and Fourier Transform of time functions.

See description in chapter 8..

7.1r.	Hilbert

Function: Hilbert Transform of Vibration signals.

Not included in this version.

7.1s.	Cepstrum

Function: Cepstrum Analysis of Vibration signals

Not included in this version.

7.1t.	Model

Function: Make the Model of a Rotor for dynamic analysis.

Method: Spreadsheet.

Limitations: None.

Use:

1. <Make> is a spreadsheet for entering data. It first asks for identification of the rotor. Then the number of elements. Then, the spreadsheet is on the screen. If there is already a rotor loaded, the data are printed on the spreadsheet. Move around with the keypad arrows. If you want to enter a value or change the existing one, type it in the appropriate place and hit <ENTER>. If you reach the right (left) of the screen, the next (previous) page appears.

2. Hit X to exit, or move the cursor vertically, out of the spreadsheet. Then, the program asks for material properties. Hit <ENTER> to retain the default value, enter a value to change it. At the end, if you answer is N, control returns to the spreadsheet.

3. Use any consistent system of units. The units of the results will depend on the input units. Use of SI throughout is suggested.

4. On the bottom of the screen, a help line is shown.

5. When you finish, plot the rotor to check the data for geometrical consistency.

8

CHAPTER EIGHT
DIAGNOSIS AND PROGNOSIS

8.1. **EXPERT** Systems for diagnosis and prognosis

Function: EXPERTS, an expert system shell for diagnosis and prognosis.

Limitations: 20 diagnoses, 30 features, 100 values.

Files needed: EXPERTS.EXE, COMPUME.SYS, EXMENUS.TXT, EXPERTS.DIR, EXPERTS.HLP.

Example system: MDOCTOR.SYM, MDOCTOR.FUR, MDOCTOR.EXP.

Method: Neural network and fuzzy set methodology is used in the development of an expert system shell aimed at developing a wide range of expert systems.

A static heteroassociative neural network module (NNM) architecture is used with separate layers of input and output neurons where the input layer is projecting on the output layers.

The input-output relation in the module is

$$a_j = \sum_{i=1}^{m} w_{ij} \phi_i \mu_A(\phi_i)$$

$$\beta_j = \theta_j(a_j)$$

where, ϕ_i is the value of the input neuron i, i=1,2,...,m

$\mu_A(\phi_i)$ is a membership function,

w_{ij} is the weight function between input i and output j,

β_j is the value of the output neuron j, j=1,2,...n,

$\theta_j(a_j)$ is a threshold function.

Knowledge is represented in this module by:

a. The weights w_{ij} of the synapses.

b. The input membership function $\mu_A(\phi_i)$

c. The output threshold values β_j

d. The network topology.

This procedure is coded in EXPERTS with the following conventions:

a. For fuzzy input quantities, the input takes values [0,1].

b. For single value input quantities, the input takes values 0 or 1.

b. For inputs which are scalar quantities, the input takes non-negative values from 0 to 1.

c. The inputs neurons are called UNITS and they are grouped in FEATURES. The names UNITS and FEATURES are default names.

The user can specify names related with the expert system under development, i.e. SYMPTOMS can replace UNITS in a diagnosis system.

d. The output neurons are called RESULTS by default, and are user definable, i.e., DIAGNOSES in a diagnosis system. They can be logic variables , true or false, or fuzzy.

e. Initially the user has to specify at least one feature, one unit in this feature and one output. The built-up of the network is done gradually to EXPERTS during the learning process. More features, units and results can be added also by the learning process.

f. The state vector includes the input vector $\{\phi\}$ and the output vector $\{\beta\}$,

$$\{v\} = \{\ \phi\ |\ \beta\ \}$$

where curly brackets indicate a transposed column vector.

The aim of the learning process is to develop and/or improve the desired mapping function between the input and output vectors. This is accomplished by adjusting the weights and the threshold values according to some learning rule.

If a result is defined for the first time, equal weight is assigned to all the units that are activated for this result. Then, the appropriate column of the weight matrix $[w_{ij}]$ is normalized to a certain value, common for all columns.

This correlational rule is expected to be improved later-on with further learning, since one can intuitively conclude that the weight of the several units does not have to be the same.

Each subsequent teaching session with the same result, and possibly different values of the input vector, is incorporated into the system by way of an error-correcting rule (Hinton, Sejnowski and Ackley 1984; Rumelhart, Hinton and Williams 1986; Sejnowski and Rosenberg 1986).

Let $\{c_{ij}\}$ the existing value of the column i of the weight matrix $[w_{ij}]$ and $\{d_{ij}\}$ the same column determined on the basis of the equal weight method but with the present value of the input vector. Then, the column i of the weight matrix is adjusted to the value

$$\{c_{ij}\}^* = \{c_i\} + \tau\{d_i\}) \qquad (4)$$

where τ_i is a correction factor which will determine the weight of the new knowledge on the existing one. Finally, vector $\{c_i\}^*$ is normalized.

During application of the error-correcting rule, new features and units might be defined during the definition of the input vector for a new case that the system is learning. The system creates the network additions and assigns zero weight to the idle synapses.

When a new unit is created, a threshold value is assigned. Each time the system learns a new case, it updates the value of the unit threshold values, if the unit values are less than the threshold ones. This is not formally correct, since threshold values should be different for the different results. It is not used at present in the interest of simplicity and memory saving and because the author feels that for engineering applications the threshold values should not depend strongly on results. This extension, however, would be straight-forward if threshold coupling is essential.

If the same value of the input vector results always to the same value of the output vector, the procedure is equivalent to the first order predicate calculus. The neural network representation allows for a fuzziness, in the form of incomplete or different inputs for a certain output. After a sufficiently large number of learning sessions, each application gives a result vector in the sense of the most frequent result corresponding to the input vector and the ones very close to it.

Many times, the values of the units, beyond their fuzziness, they are uncertain, in the sense that either they are based on incomplete or uncertain information which might or might not affect substantially the result vector. In the neural network formulation, this sensitivity can be quantified with the Euclidean distance

$$d_j = \{\Sigma[\beta(\phi+D\phi_j)-\beta(\phi)]^2\}^{1/2} \qquad (5)$$

The system computes the Euclidean distance for variations $D\phi_j$ due to the change in value of any unit j.

Available experience for failure diagnosis in turbomachinery (Shore 1980) was utilized to initially teach the system. Additional diagnoses from the user's experience can be taught to the system and additional features and diagnoses defined.

It is expected that as is the system will not lead to unambiguous diagnoses for particular fields of applications. Repeated utilization and learning from application experience can lead to an adoptive orientation to a particular field of application.

Use:

DEFINITIONS

Some definitions are first in order:

1. **CAUSE** is the objective of the diagnosis. The user specifies a number of possible causes. For example, a doctor making an Expert System for medical diagnosis will specify different causes of not feeling well, such as FLU or CANCER or COLD or ULCER,etc. An engineer making an Expert System for Machinery Fault Diagnosis would specify causes as UNBALANCE, MISALIGNMENT, WEAR, etc.

2. **FEATURES** are the characteristics which will be looked upon for the diagnosis of the CAUSE of the problem. For example, in the Doctor's system, features are the FEVER, the PAIN, etc. In the engineer's problem, features are the VIBRATION, NOISE, TEMPERATURE, etc.

3. **SYMPTOMS** are the different possible values of the features, i.e. for the feature FEVER, symptoms can be NORMAL, HIGH, BETWEEN 100-103, etc. For the feature NOISE the symptoms can be HIGH PITCH, ABOVE 100 db, etc. Up-to 10 symptoms per feature can be specified.

CREATION OF A NEW EXPERT SYSTEM

To create a new expert system, one has to start with the use of the 'Create System' selection. The program will ask for a name (up-to 8 characters) and a Descriptive Title (up-to 25 characters). Further, the program will prompt for the definition of, at least, one cause, one feature and one symptom of this feature. Of course, the user will want to specify more causes, features and symptoms to have a meaningful system.

However, one does not have to do it from the beginning. He can add later on as many causes, features and symptoms as needed at any time using the 'AddFeatures' selection.

UPDATING AN OLD EXPERT SYSTEM

Intelligence is an important feature of EXPERTS. The Expert Systems it generates can learn by adopting any new diagnosis that you verified.

Using the 'Update Experience' selection you can teach your Expert System the new diagnosis. In this way, you can gradually teach your system and make it a true expert. This is important in many cases where precast expert systems cannot function because they are not necessarily based on experiences on similar cases. Or, one can take an existing Expert System, made with EXPERTS, and enhance it gradually, just about as the physical learning process. Moreover one does not have to write lengthy and complicated rules. EXPERTS constructs its rules from the experience when you teach it. Every time you teach it a new diagnosis, it creates the appropriate rule.

DIAGNOSIS

The program checks the symptoms against the current input values and makes diagnosis by inference.

The diagnosis might not be unique. The different causes are listed in descending order of a parameter which signifies the relative weight of each cause, proportional to the number of cases where the specified symptoms yielded the respective diagnosis. Many times identical symptoms lead to different diagnoses. EXPERTS assigns a frequency of occurrence depending on the number of cases it was taught. On this basis, EXPERTS many times does not give one diagnosis but ranks several ones on the frequency-of-occurrence basis, on the 0 to 100 scale. If the diagnoses you taught the system are unique, then the diagnosis will yield one 100 and all the other causes 0. The diagnosis will then be unique.

HOW TO GET HELP

EXPERTS menu page has a bottom line help describing the current main menu selection at the top line. This menu is selected with the horizontal keypad arrows or by hitting the capital letter indicated in the menu line.

To get help on any sub-menu item in the sub-menu rectangles, select first the submenu item using the keypad arrows and then hit F1.

MENU SELECTIONS

QUIT: Quits EXPERTS and returns to the root menu.

Upon return to root menu you do not lose the current case since it is saved in the Symptom File LASTCASE.LST where LASTCASE is the name of the last Expert System you have loaded.

Directory, Expert Systems: Directory of all Expert Systems created by EXPERTS.

Prints a directory of all Expert System created previously by EXPERTS on the current drive in yellow and the help line which explains each Expert System name in green.

Directory, Symptom Files: Directory of all Symptom vectors used by Expert System LASTCASE.

Prints a directory of all Symptom Vectors saved previously having the extension *.SYM on the current drive. You can obtain a more general directory with the LoadFile selection.

Systems, Load: Loads an already created Expert System.

Loads an Expert System from the ones already created on the current drive. Use Directory before to make sure that your Expert System is on the current drive. If you have already loaded another expert system, it is replaced in memory but it remains on the current drive.

Systems, Create: Creates a new Expert System on the current drive. An Expert System should have the following components:

1. At least one CAUSE of problem to be diagnosed, ie FLU, ULCER,...

2. At least one diagnostic FEATURE, that is a category of symptoms on which a diagnosis will be based, such as FEVER, PAIN,...

3. At least one SYMPTOM of the above feature, such as NORMAL, HIGH. A maximum of 10 Symptoms per feature is allowed.

Maximum total number of CAUSES, FEATURES, SYMPTOMS is specified in the EXPERTS.SYS file.

Systems, Remove: Removes an already created Expert System from the directory.

Removes an already created Expert System from the directory from the current drive. Use Directory before to make sure that your Expert System is on the current drive. Upon removal, the system files will not be erased from memory but they remain there, with the name EXFILE.* where EXFILE the Expert System name. To reinstate the system, you add its name EXFILE and an explanation line in the file EXPERTS.DIR using an ASCII text editor.

Symptoms, Load: Load a Symptom File with a previously saved case.

Loads a previously saved file with the symptoms of a case. Use Wild Character * for a directory of files with a path which you will specify:

Enter File Name: C:\exp*.SYM

will print a list of all files in the directory c:\exp with the extension .SYM, as with DOS conventions.

Symptoms, Save: Save current case on a Data File

Saves a file with the symptoms of the current case for later use Use Wild Character * for a directory of existing files with a path which you will specify:

Enter File Name: C:\exp*.SYM

will print a list of all files in the directory c:\exp with the extension .SYM, as with DOS convensions.

Symptoms, Input Symptoms: Symptoms for a new case or modify current case

You will enter here the symptoms for any number of features you want or you will modify the current values. The program starts with some default values, if you do not load a previously saved case. Every line corresponds to one feature. On the lower part of the screen, the symptoms for the current feature are printed. You have to select one by typing the corresponding number, or none by typing a space. Then, move vertically with the keypad arrows. You can move in any direction you want, up or down, and modify the data in any sequence. When you exit the features, the program displays the next 10 features until all features have been shown on the screen. To continue, either exit vertically or hit X. The program prompts you to answer if the data are correct. If your answer is N (no) the input procedure is repeated.

A word of caution: If you type a value, hit an arrow in any direction before you hit X so that the entry will be registered.

Diagnosis: Perform diagnosis

The program checks the symptoms against the current input values and performs a diagnosis. Before you use this option, you must either load a previous case or use the input selection to specify or change the symptoms. Otherwise the program will use the built-in values which are valid only for the default problem.

The diagnosis might not be unique. The different causes are listed in descending order of a parameter which signifies the relative weight of each cause, proportional to the number of cases where the specified symptoms yielded the respective diagnosis.

Diagnosis, Sensitivity: Perform Sensitivity Analysis of the current case

When we specify a symptom, there is uncertainty in most cases. In this case, Sensitivity Analysis helps identify what change in the diagnosis is caused by changing anyone of our selection of symptoms. The input lines return on the screen, as described in INPUT SYMPTOMS above. You can change any ONE input. Upon hitting a keypad arrow key, the system performs diagnosis again with the new symptom. For comparison, the previous diagnosis is desplayed on the right of the screen.

Diagnosis, Show Rule: Shows the features and symptoms of last diagnosis.

The features and symptoms specified which led to the last diagnosis are displayed in the rule-form IF .... AND ... AND... ...THEN.....

Extend, Update Experience: Incorporates users experience into the Expert System

This is a learning feature. Any diagnosis, wright or wrong, can be compared with the eventual identification of the problem. The true diagnosis can be incorporated into the Expert System. In this way the system can be taught gradually expertise in any area. The system can start from scratch, that is without previous experience, and then gradually be educated by the user based on his experience.

Extend, Add Features: Incorporates new diagnostic FEATURES and CAUSES of the problem

It incorporates new diagnostic FEATURES and CAUSES of the problem into the Expert System as the user becomes more familiar with the problem. Therefore, the development of the Expert System can start from an elementary form and gradually be improved by adding new diagnostic features and problem causes.

Probable Errors: Absence of system files.

Probable Errors: Absence of system files.

```
 Quit   Direct System sYmpt  diaGn  Extend Util        COMPUME           EXPERTS
```

MDOCTOR DIAGNOSIS:

Rank	Type	Score%	Source of Problem
1	33	100	Cracked Rotor
2	2	80	Permanent bow or lost parts
3	1	80	Initial Unbalance
4	11	70	Journal and bearing eccentricity
5	12	60	Bearing damage
6	6	60	Foundation distortion
7	5	60	Permanent casing distortion
8	29	60	Pressure pulsations
9	27	50	Structural resonance of support
10	26	50	Structural resonance of casing
11	25	50	Overhang critical
12	24	50	Coupling critical
13	23	50	Rotor & bearing system critical
14	14	50	Bearing unequal stiffness
15	28	50	Structural resonance of foundation
16	7	50	Seal rub
17	4	50	Temporary casing distortion

Hit any key to return to the Main Menu, M for more

8.2. ANALYZER

Function: A Signal Analysis program.

Files needed: ANALYZER, one or more FFT.* files.

Hardware required: Graphics monitor card, 640 kB memory.

Method: The discrete Fourier Transform of the periodic function harmonic functions g(t) is defined as

$$g(t) = \sum_{n=-\infty}^{\infty} G_n e^{in_\omega t} \qquad (a)$$

$$G_n = 1/T \int_{0}^{T} g(t)\, e^{-in_\omega t}\, dt \qquad (b)$$

For $T \to \infty$, the spacing 1/T between the harmonics tends to zero and G becomes a continuous function of $f = \omega/2\pi$. Then,

$$G(f) = \int_{-\infty}^{\infty} g(t)e^{-i2\pi ft}\, dt$$

$$g(t) = \int_{-\infty}^{\infty} G(f)e^{i2\pi ft}\, df$$

If truncation is performed in the time series and discretization and truncation is similarly performed in the frequency function, the discrete transform becomes

$$G_k = \sum_{n=0}^{N-1} g_n e^{-i(2\pi kn/N)}$$

$$g_n = (1/N) \sum_{k=0}^{N-1} G_k\, e^{i(2\pi kn/N)}$$

A calculation procedure, known as the *Fast Fourier Transform* or *FFT algorithm*, obtains the same result with a much smaller number of complex multiplications: One way of expressing the transform with the following matrix equation:

$$\{G\} = (1/N)[A]\{g\}$$

where

   {G} is a column vector representing the N complex frequency components G_k, k=0,1,2,...N-1

   N is the number of the equal time intervals between samples.

[A] is an NxN square matrix of unit vectors which depend only on the number of samples, $\quad a_{kn} = e^{-i(2\pi kn/N)}$

{g} is a column vector representing the N time samples g_n, n=0,1,2,..N-1

Use: You select ANALYZER from the COMPUME menu or type ANALYZER in your COMPUMECH subdirectory. The simulated Spectrum Analyzer and the Tape Recorder appear on the screen. The control buttons can be depressed by hitting the keyboard keys with the same starting letter on the button. Depressing X lights the button and transfers control to the tape recorder. Pressing the + button (+ of the keypad) advances the channel and pressing the - button (- of the keypad) retracts the channel number. The tape recorder reads the data file FFT.N, where N is the channel number selected. We set channel number to 1 and hit ENTER. It takes a few seconds to read the file with the time signal. The file window on the screen shows the number of samples, the time step and the frequency resolution.

When the reels stop, control returns to the control console. We press the T key to view the time function.

Pressing the F key invokes the FFT analysis. After a few seconds, the light on the F button is off. We can now plot the results by pressing the A key for amplitude.

To zoom on this frequency, the Z key is pressed. The program asks for the frequency range. We specify 0<f<100 Hz.

Pressing the D key invokes the DFT analysis. On the lower window, the program asks for the desired vibration range. After a few seconds, the light on the D button is off. We can now plot the results by pressing the A key for amplitude. The discrete spectrum appears on the screen.

Example: A vibration signal was digitized and saved in the FFT.1 file. Perform FFT and DFT analysis and find the dominant frequencies.

ANALYZER was selected first, then channel 1 (FFT.1) was retrieved with the X command. The time signal, which with the zoomed FFT and DFT spectra are shown below. 16, 32 and 48 hz frequencies are apparent in the spectra.

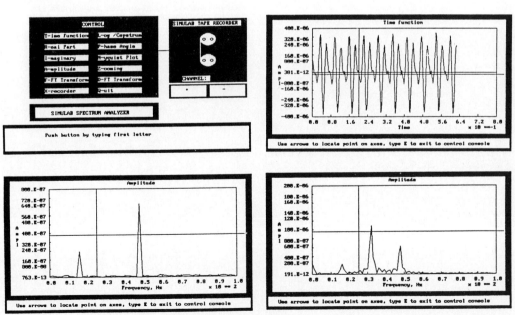

9

CHAPTER NINE
MATERIALS

9.1. **MATERIALs** Database

Function: Retrieves material properties from a database, draws Goodman diagrams and computes the safety factor for combined static and dynamic loading. Uses the materials files with extension .MTL and a pointer file POINTER.TXT .

Files needed: MATERIAL.EXE, POINTER.TXT, MAT1.MTL to MAT8.MTL, MAT13.MTL, MAT16.MTL, CARBON.MTL, COMPUME.SYS.

Reference: Dimarogonas 1988, chapter 7.

Hardware Requirements: 512k, 1 FD 360k, EGA card, monochrome monitor.

Limitations: None

Method: A pointer file organizes the material files *.MTL. Each category of material occupies two lines. The first has the name of material and an explanation. The second line the number of files, the name of files, the number of materials, their number in the respective material filefile, ends with EL,. New materials and files can be added by the user.

```
  ALUMINUM                      , ALUMINUM
, 0, EL,
  WROUGHT AL ALLOYS             , WROUGHT AL ALLOYS
, 0, EL,
  CAST AL ALLOYS                , CAST AL ALLOYS
, 0, EL,
  STRUCTURAL STEELS             , STRUCTURAL STEEL
, 1, MAT2.MTL, 7 , 1 ,2 ,3,4,5,6,7,EL,
  CARBON STEELS                 , ALLOY DIRECT HARDENING STEELS
, 1, CARBON.MTL , 11, 1 ,2 ,3,4,5,6,7,8,9,10,11,EL,
  ALLOY STEELS                  , ALLOY DIRECT HARDENING STEELS
, 1, MAT3.MTL , 13, 1 ,2 ,3,4,5,6,7,8,9,10,11,12,13,EL,
  STAINLESS STEELS              , STAINLESS FERRITIC/MARTENSITIC
, 1, MAT4.MTL , 12, 1 ,2 ,3,4,5,6,7,8,9,10,11,12,EL,
  HEAT RESISTANCE STEELS        , FLAME AND INDUCTION HARDENING
, 1, MAT5.MTL , 11, 1 ,2 ,3,4,5,6,7,8,9,10,11,EL,
  COPPER                        , COPPER
, 1, MAT6.MTL, 1 , 9 ,EL,
```

```
  BRONZE                          , CU-SN BRONZE
, 1, MAT6.MTL, 1 , 10,EL,
  BRASS                           , WROUGHT CU-ZN BRASS#CAST CU-ZN BRASS
, 1, MAT6.MTL, 2 , 13,14,EL,
  CAST IRON                       , AUSTENITIC CAST IRON
, 1, MAT7.MTL , 18, 1 ,2 ,3,4,5,6,7,8,9,10,11,12,13,14,15,16,17,18,EL,
  CAST STEEL                      , MALL IRON/BL#MALL IRON/PERL#STEEL CASTINGS
, 1, MAT8.MTL, 30, 1 ,2
,3,4,5,6,7,8,9,10,11,12,13,14,15,16,17,18,19,20,21,22,23,24,25,26,27,28,29,30
,EL,
  MG ALLOYS                       , CAST MG ALLOYS#WROUGHT MG ALLOYS
, 1, MAT6.MTL, 2 , 4 ,5 ,EL,
  TITANIUM ALLOYS                 , TITANIUM ALLOYS
, 1, MAT6.MTL, 2, 2 ,3 ,EL,
  AL-MG ALLOYS                    , AL-MG ALLOYS
, 1, MAT6.MTL, 1 , 3 ,EL,
  CAST MG ALLOYS                  , CAST MG ALLOYS
, 1, MAT6.MTL, 1 , 4 ,EL,
  WROUGHT MG ALLOYS               , WROUGHT MG ALLOYS
, 1, MAT6.MTL, 1 , 5 ,EL,
  AL-ZN-MG ALLOYS                 , AL-ZN-MG ALLOYS
, 1, MAT6.MTL, 1 , 6 ,EL,
  CAST AL-TI-CU ALLOYS            , CAST AL-TI-CU ALLOYS
, 1, MAT6.MTL, 1 , 7 ,EL,
  CAST AL-SI-MG ALLOYS            , CAST AL-SI-MG ALLOYS
, 1, MAT6.MTL, 1 , 8 ,EL,
  CU-NI ALLOYS                    , CU-NI ALLOYS
, 1, MAT6.MTL, 1 , 11,EL,
  CU-AL ALLOYS                    , CU-AL ALLOYS
, 1, MAT6.MTL, 1 , 12,EL,
  ACETAL                          , ACETAL
, 1, MAT6.MTL, 1 , 15,EL,
  ACETAL GLASS REINFORCED         , ACETAL GLASS REINFORCED
, 1, MAT6.MTL, 1 , 16,EL,
  NYLON 6/12                      , NYLON 6/12
, 1, MAT6.MTL, 1 , 17,EL,
  NYLON GLASS REINFORCED          , NYLON GLASS REINFORCED
, 1, MAT6.MTL, 1 , 18,EL,
  POLYESTER                       , POLYESTER
, 1, MAT6.MTL, 1 , 19,EL,
  POLYESTER GLASS REINFORCED, POLYESTER GLASS REINFORCED
, 1, MAT6.MTL, 1 , 20,EL
```

File MAT1.MTL contains the physical properties for each of the materials listed:

Specific Weight γ

Young's Modulus E, GPa

Shear Modulus G, GPa

Thermal expansion coefficient $10^6\alpha$, $/^{\circ}C$

Specific heat c, kJ/kg$^{\circ}$C

Thermal conductivity k, kJ/mh$^{\circ}$C

Electrical resistance ρ, $\mu\Omega$m

Poisson Ratio ν

More materials can be added by the user. A printout of the file MAT1.MTL follows:

```
"ALUMINUM","2.70","62.1","23.3","22.2","0.921","775","0.027","0.34"
"WROUGHT AL ALLOYS","2.72","74","28","22","0.921","500","0.045","0.34"
"CAST AL ALLOYS","2.7","68","28","23","0.921","560","0.053","0.34"
"STRUCTURAL STEELS","7.85","210","85","11.45","0.477","190","0.17","0.27"
"ALLOY STEELS","7.85","210","84","11.4","0.51","120","0.7","0.27"
"STAINLESS STEELS","7.7","200","86","18","0.5","45","0.7","0.29"
"HEAT RESISTANCE STEELS","7.83","210","82","11.45","400","45","0.8","0.35"
"COPPER","8.97","117","50","16","0.385","1400","0.017","0.295"
"BRONZE","8.5","112","41","17","0.385","600","0.045","0.295"
"BRASS","8.5","109","40","17","0.377","245","0.08","0.295"
"CAST IRON","7.5","66 - 170","9.6 - 28","10","0.586","180","0.9","0.2"
"CAST STEEL","7.83","207","77","12.5","0.48","134","1","0.31"
"CARBON STEEL","7.83","207","77","12.5","0.48","134","1","0.31"
"MG ALLOYS","1.8","45","16.6","26","1.05","300","0.14","0.3"
"TITANIUM","4.51","107","41","8.5","0.469","50","0.12","0.34"
"","","","","","","","",""
```

Each of the material files contains:

> The material number,
>
> the commercial name,
>
> the ISO designation,
>
> the USA designation,
>
> the USA standard,
>
> the DIN designation,
> the ultimate strength S_u, MPa,
>
> the yield strength S_y, MPa,
>
> the fatigue strength S_n, MPa,
>
> the elongation at fracture %,
>
> application notes.

The program computes the safety factor if the general state of stress is given for the material selected. The Goodman diagram is used for safety factor [Dimarogonas 1988]. The assymetry factor is first defined:

$$r_D = (S_e/S_u)[(S_u - S_y)/(S_y - S_e)], \quad r = \sigma_m/\sigma_r$$

For $r > r_D$, $N = S_u/(\sigma_m + \sigma_r S_u/S_e)$,

For $r < r_D$, $N = S_y/(\sigma_m + \sigma_r)$.

where σ_m the mean stress, σ_r the stress range and S_e the effective fatigue strength.

Use: To use the program (User types what below is underlined):

Select from MENUMEC or type <u>MATERIAL</u>

The first page has the main menu. List of all available materials with their commercial names, properties of a specific material or the materials which have a chosen property within specified limits. In both cases, you can ask for the Goodman diagram. The program asks then for all static and alternating stresses and computes the safety factor from the Goodman diagram.

```
┌──────────────────────────────────────────────────────────────────────┐
│              COMPUTER AIDED MACHINE DESIGN: MATERIALS DATABASE          │
├──────────────────────────────────────────────────────────────────────┤
│                                                                        │
│                                                                        │
│                                                                        │
│              1. Listing of available Materials                         │
│              2. Properties of a Specific Material                      │
│              3. Materials with a Property in a given Range             │
│              4. QUIT...Return to CAMD Menu                              │
│                                                                        │
│                                                                        │
│                                                                        │
├──────────────────────────────────────────────────────────────────────┤
│ ENTER YOUR SELECTION:                                                  │
└──────────────────────────────────────────────────────────────────────┘
```

```
┌──────────────────────────────────────────────────────────────────────┐
│                     LIST OF MATERIALS AVAILABLE                        │
├──────────────────────────────────────────────────────────────────────┤
│   1 . ALUMINUM                    16 . CAST MG ALLOYS                  │
│   2 . WROUGHT AL ALLOYS           17 . WROUGHT MG ALLOYS               │
│   3 . CAST AL ALLOYS              18 . AL-ZN-MG ALLOYS                 │
│   4 . STRUCTURAL STEELS           19 . CAST AL-TI-CU ALLOYS            │
│   5 . ALLOY STEELS                20 . CAST AL-SI-MG ALLOYS            │
│   6 . STAINLESS STEELS            21 . CU-NI ALLOYS                    │
│   7 . HEAT RESISTANCE STEELS      22 . CU-AL ALLOYS                    │
│   8 . COPPER                      23 . ACETAL                          │
│   9 . BRONZE                      24 . ACETAL GLASS REINFORCED         │
│  10 . BRASS                       25 . NYLON 6/12                      │
│  11 . CAST IRON                   26 . NYLON GLASS REINFORCED          │
│  12 . CAST STEEL                  27 . POLYESTER                       │
│  13 . MG ALLOYS                   28 . POLYESTER GLASS REINFORCED      │
│  14 . TITANIUM ALLOYS                                                  │
│  15 . AL-MG ALLOYS                                                     │
├──────────────────────────────────────────────────────────────────────┤
│ Hit ENTER to continue                                                 │
└──────────────────────────────────────────────────────────────────────┘
```

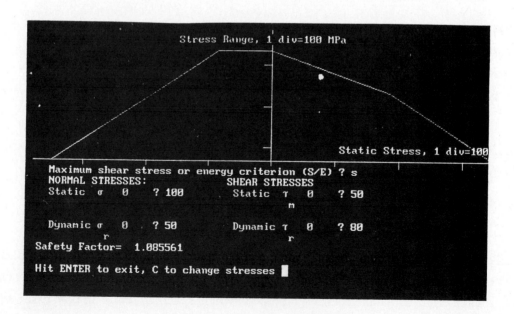

Probable errors: Bad file, error in the data files. File missing.

9.2. SECTIONS

Function: Properties of complex sections.

Reference: Dimarogonas 1988, chapter 8.

Files needed: SECTIONS.EXE, COMPUME.SYS, a SECTION*.DAT file.

Hardware Requirements: 512k, 1 FD 360k, EGA card, monochrome monitor.

Limitations: Up to 20 polygons.

Method: The section is described by a number of polygons. The solid sections are traced clockwise, the openings (holes) counter-clockwise. The coordinates of the polygon node i are (x_i, y_i) in a Cartesian coordinate system.

The section area is

$$A = \Sigma(x_i - x_{i-1})(y_i + y_{i+1})$$

The coordinates of the area center are

$$y_G = \Sigma(x_i - x_{i+1})(y_i^2 + y^2_{i+1} + y_i y_{i+1})/6A$$

$$x_G = \Sigma(y_{i+1} - y_i)(x_i^2 + x^2_{i+1} + x_i x_{i+1})/6A$$

The moments of inertia:

$$Ixx_G = \Sigma(x_i - x_{i+1})(y_i^3 + y_i^2 y_{i+1} + y_i y_{i+1}^2 + y^3_{i+1})/12$$

$$Iyy_G = \Sigma(y_{i+1} - y_i)(x_i^3 + x_i^2 x_{i+1} + x_i x_{i+1}^2 + x^3_{i+1})/12$$

$$Ixy_G = \Sigma(x_i - x_{i+1})[x_i(9y_i^2 + 6y_i y_{i+1} + 3y^2_{i+1}) + x_{i+1}(3y_i^2 + 6y_i y_{i+1} + 9y^2_{i+1})$$

The pricipal stresses are

$$I_{1,2} = (I_{xx} + I_{yy}) \pm 0.5[(I_{xx} + I_{yy})^2 + 4I^2_{xy}]^{1/2}$$

and their angle in respect to the coordinate system (x, y) is

$$\tan2\alpha = 2I_{xy}/(I_{yy} - I_{xx})$$

Use:

To use the program (User types what below is underlined):

1. Using the SOLID program, on the x,y plane define the section as polygons traced CW. Holes are also polygons traced CCW. Save the polygons in a data file. The data file can also be made in a text editor. For, example, file SECTION1.DAT is

```
NODE,  10 ,  10 ,  10
NODE,  40 ,  10 ,  10
NODE,  40 ,  17 ,  10
NODE,  28 ,  17 ,  10
NODE,  28 ,  72 ,  10
NODE,  40 ,  72 ,  10
NODE,  40 ,  78 ,  10
NODE,  10 ,  78 ,  10
```

```
NODE,  10 , 72 , 10
NODE,  22 , 72 , 10
NODE,  22 , 18 , 10
NODE,  10 , 18 , 10
NODE,  10 , 10 , 10
POLYGON, 13 , 1 , 2 , 3 , 4 , 5 , 6 , 7 , 8 , 9 , 10 , 11 , 12 , 1
SOLEND
```

First, the nodes are defined by their x,y,z coordinates. The latter is irrelevant and can be any number, but it must be there.

For every polygon, a node connectivity line should be added:

POLYGON, number of nodes, n_1, n_2, ..., n_n.

The file ends with the SOLEND delimiter.

Select SECTIONS from MENUMEC or type SECTIONS.

The first window is for program identification. It asks for the name of the file prepared with SOLID. Enter it at the ? prompt.

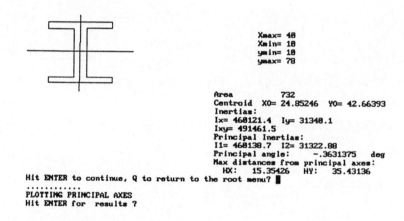

```
                            Xmax= 48
                            Xmin= 18
                            ymin= 18
                            ymax= 78


Area              732
Centroid  XO= 24.85246  YO= 42.66393
Inertias:
Ix= 468121.4  Iy= 31348.1
Ixy= 491461.5
Principal Inertias:
I1= 468138.7  I2= 31322.88
Principal angle:    -.3631375    deg
Max distances from principal axes:
    HX:    15.35426   HY:    35.43136

Hit ENTER to continue, Q to return to the root menu? █
.............
PLOTTING PRINCIPAL AXES
Hit ENTER for  results ?
```

In creating the section, solid sections are traced clockwise, holes counterclockwise. Any number of polygons can be used in one SOLID file, to form a compound section.

Next is then the results window.

The program plots the section, the principal axes and the section properties.

Probable errors: File error. Bad data file or file not in the current directory.

9.3. FAILURE

Function: Analysis of failure under complex state of stress or combined loads.

Files needed: FAILURE.EXE, COMPUME.SYS.

Reference: Dimarogonas 1988.

Hardware Requirements: 512k, 1 FD 360k, any, monitor.

Limitations: None

Method: The Jacobi method is used to find the principal stresses. The general state of stress is expressed by the matrix

$$[A] = \begin{bmatrix} \sigma_{xx} & \sigma_{xy} & \sigma_{xz} \\ \sigma_{yx} & \sigma_{yy} & \sigma_{yz} \\ \sigma_{zx} & \sigma_{zy} & \sigma_{zz} \end{bmatrix}$$

Assume the eigenvalue problem in the form

$$-\lambda \underline{I}x + \underline{A}x = \underline{0}$$

Define a rotation matrix $\underline{R}$ for i,j = 1, 2, 3, with angle of rotation

$$\tan 2\theta = 2a_{ij}/(a_{ii}-a_{jj}).$$

For example, for i=1, j=3

$$R = \begin{bmatrix} \cos\theta & 0 & \sin\theta \\ 0 & 1 & 0 \\ -\sin\theta & 1 & \cos\theta \end{bmatrix}$$

Iterate

$$A^* = R^T A R$$
$$R^* = R^* R$$

until the off diagonal elements of $\underline{A}^*$ become less than a prescribed number. Then, the principal stresses are the diagonal elements of $\underline{A}^*$ and the direction cosines of the principal directions are the columns of the matrix $\underline{R}^*$.

The equivalent stresses are computed with 3 methods:

a. The maximum shear stress theory

$$\sigma_{eq-max} = \max|\sigma_i - \sigma_j|, \; i,j = 1,2,3$$

b. The distortion energy theory

$$\sigma_{eq-max} = \{1/2\,[(\sigma_1 - \sigma_2)^2 + (\sigma_2 - \sigma_3)^2 + (\sigma_3 - \sigma_1)^2]\}^{1/2}, \; i,j = 1,2,3$$

c. The limiting state of stress theory

$$\sigma_{eq-max} = \max|\sigma_i - k\sigma_j|, \; i,j = 1,2,3$$

where k = S$_{ut}$/S$_{ct}$, the ratio of the tensile to compressive stress.

Use: To use the program (User types what below is underlined):

Select from MENUMEC or type FAILURE.

The first window is for program identification and data.

Enter first the three normal stresses σ_{xx}, σ_{yy}, σ_{zz}.

```
Enter NORMAL stress Sxx < 1866.026 >? 1000
Enter NORMAL stress Syy < 133.9746 >? 500
Enter NORMAL stress Szz < 2000 >? 100
Enter SHEAR stress Txy < 0 >? 600
Enter SHEAR stress Txz < 0 >?
Enter SHEAR stress Tyz < 0 >?
```

```
RESULTS:
-----------------

----------Principal stresses----------
   100.0E+00    100.0E+00    140.0E+01
-------------eigenvectors----------------
     0.0E+00    554.7E-03    832.1E-03
     0.0E+00   -832.1E-03    554.7E-03
   100.0E-02      0.0E+00      0.0E+00

Do you want the limiting states of stress theory? (Y/N) ? y

Enter ratio k=Sut/Suc                     ? .5


----------Equivalent stresses------------

Maximum shear stress theory,             Seq= 1300
Distortion strain energy theory,         Seq= 1300
Limiting states of stress theory (Mohr),Seq= 1350
Hit ENTER to continue, Q to return to the root menu?
```

Enter then the three shear stresses σ_{xy}, σ_{xz}, σ_{yz}.

The program prints the principal stresses σ_1, σ_2, σ_3 and the eigenvalues which are the direction cosines of the principal planes. Finally, the program prints the equivalent stresses with the maximum shear stress theory, the distortion shear energy theory and the Mohr theory of limiting stress.

9.4. Reliability and **SAF**ety **FAC**tor

Function: Statistical evaluation of the safety factor.

Reference: Dimarogonas 1988.

Hardware Requirements: 512k, 1 FD 360k, EGA, color monitor.

Limitations: None.

Method: Given are the reliability A, the mean load $\underline{L}/\underline{S}_L$ and mean capacity $\underline{c}/\underline{S}_c$ ratios, where $\underline{L}$ and $\underline{c}$ the mean values and $\underline{S}_L$ the standard deviations. The safety factor will be the solution od the equation

$$1 - A = \int_0^\infty \frac{e^{(-t_2/2)}}{(2\pi)^{1/2}} \, dt, \quad t = \frac{1 - N}{[(\underline{L}/\underline{S}_L)^2 + (\underline{c}/\underline{S}_c)^2]^{1/2}}$$

assuming standard distribution for the load and the capacity (strength). The program plots the distributions for every evaluation of the safety factor.

Use: To use the program (User types what below is underlined):

Select from MENUMEC or type SAFFAC.

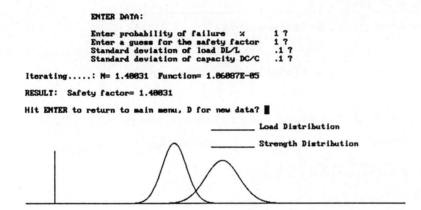

```
ENTER DATA:

Enter probability of failure   x      1 ?
Enter a guess for the safety factor   1 ?
Standard deviation of  load DL/L      .1 ?
Standard deviation of capacity DC/C   .1 ?

Iterating.....: N= 1.40031  Function= 1.06007E-05

RESULT:  Safety factor= 1.40031

Hit ENTER to return to main menu, D for new data? █
```

_____ Load Distribution
_____ Strength Distribution

Input-output are shown above. A safe guess for the safety factor is 1. Try different values >1 if there is no convergence.

Example: An example of a safety factor computation is built-in into the program. Select input file SAFFAC.DAT. Hit ENTER repeatedly and the example will be executed.

10

CHAPTER TEN
MECHANICAL DESIGN

10.1. **FITS** and tolerances

Function: Computation of limit dimensions of fits.

References: Dimarogonas 1988, ISO 286, American National Standards Institute, Washington, D.C.

Hardware Requirements: 256k, 1 FD 360k, any card, monitor.

Files needed: FITS.EXE

Input: Nominal Diameter d, shaft quality it, hole quality IT, category for basic hole system.

Output: Limit dimensions of shaft d_{upper}, d_{lower} and hole D_{upper}, D_{lower}.

Method: According to ISO 280, for basic hole system.

	Free fit	Shrink fit
Shaft:	$d_{max} = d-a_o$, $d_{min} = d-a_o-it$	$d_{max} = d-a_u+IT$, $d_{min} = d-a_u+IT+it$
Hole:	$D_{min} = d$, $D_{max} = d+IT$	$D_{min} = d$, $D_{max} = d+IT$

where d the nominal diameter, it the shaft tolerance, IT the hole tolerance, a_o the minimum clearance, a_u the minimum interference.

Use: To use the program (User types what below is underlined):

To use the compiled program, select from MENUMEC or type <u>FITS</u> from DOS.

The program computes the limit dimensions **in the basic hole system**.

```
COMPUTER AIDED MACHINE DESIGN

Problem ............ Default example      ?
Enter Basic Dimension (mm).......... 80   ?
Enter Quality of Shaft  it ......... 7    ?
enter Quality of Hole   IT ......... 8    ?
Enter category of Shaft (Hole-H) ....g    ?


   80 H 8 /g 7   fit


SHAFT:   79.95789  <  d  <  79.98891
HOLE:    80  < D <   80.04848
emax:   -1.109151E-02
emin:   -9.059015E-02

Hit ENTER to change fit, P for another problem, Q to quit?
```

The user enters nominal diameter (mm), shaft and hole quality (6-12), shaft category. It yields limit dimensions and limits of the clearance (negative) or interference (positive).

Example: An example is built-in into the program. Hit ENTER repeatedly and the example will be executed.

Possible Error Messages: Categories or qualities out of range.

10.2. OPTIMUM

Function: Minimum of a function, subject to equality and inequality constraints.

Files needed: OPTIMUM.OBJ and OPTIFUN.OBJ

Reference: Dimarogonas, A.D. 1988.

Hardware Requirements: 512k, 1 FD 360k, EGA card, monochrome monitor.

Limitations: Up to 10 variables. Limits can be extended by appropriate changes in the program.

Method: A penalty function is defined as

$$F(x_1, x_2, \ldots, x_n) =$$
$$= f(x_1, x_2, \ldots, x_n) + K_1 \Sigma [g_i(x_1, x_2, \ldots, x_n)]^2 +$$
$$+ K_2 \Sigma <h_i(x_1, x_2, \ldots, x_n)>^2$$

where $f(x_1, x_2, \ldots, x_n)$ is an objective function to be minimized

$\qquad g_i(x_1, x_2, \ldots, x_n)$, $i = 1, 2, \ldots,$ m are equality constraints

$\qquad h_i(x_1, x_2, \ldots, x_n)$, $i = 1, 2, \ldots,$ k are inequality constraints

$\qquad <y> = y$ if $y>0$, zero otherwise

$\qquad x_1, x_2, \ldots, x_n$ are the design parameters.

Given an initial state $x_{01}, x_{02}, \ldots, x_{0n}$, an increment of the design parameters $\Delta x_1, \Delta x_2, \ldots, \Delta x_n$ is computed with several methods, so that the value of F will become smaller at each step until a minimum is reached. K_1 and K_2 are constant weighting factors for the constraints.

Use: An .EXE file cannot be used in this case diretly. Go to the source code OPTIFUN.HLL, where HLL is the MICROSOFT igh level langauage of your choise, BAS, FOR, C for QuickBasic, Fortran and C respectively. Define in the subroutine **UserPenalty** your penalty function. To this end, the program uses wx(1), wx(2),... as variable names and expects the penalty function to be named wF, the equality constraints wG1, wG2,... and the inequality constraints wH1, wH2,... Use any parameter names, but preferably start all your variables with w to avoid conflict with program parameters in COMMON. The equality multiplier should be wK1 and the inequality multiplier should be wK2. You can use as many equality and inequality constraints you want but only 4 of eacxh will be reported on the screen.

Example: For example, suppose that a vertical cylindrical water container is to be optimized so that the weight of the container is minimum while the volume is given as $V = 14$ m^3, the floor space should be no more than 2 m^2 and the material strength on the basis of the peripheral stress pr/t is 10 MPa. There are three variables: Diameter d, height H, thickness T . The first three are expected to be several meters each, the thickness several mm. The variable definition should be:

Container material to be minimized $f = (\pi dH + 2\pi d^2/4)T$

Equality constraint $g1 = pr/T - 10 \times 10^6 = 9810 Hd/2T - 10 \times 10^6 = 0$

Inequality constraint $h1 = 2\,m^2 - \pi d^2/4 > 0$

Define in the subroutine **UserPenalty** your penalty function. To this end, the program uses wx(1), wx(2),... as variable names and expects the penalty function to be named wF, the equality constraints wG1, wG2,... and the inequality constraints wH1, wH2,... Use any parameter names, but preferably start all your variables with w to avoid conflict with program parameters in COMMON. The equality multiplier should be wK1 and the inequality multiplier should be wK2. You can use as many equality and inequality constraints you want but only 4 of eacxh REM User defined penalty function for OPTIMUM

To use OPTIMUM:

  1. Prepare below the penalty function definition. The example below is in
     Microsoft QBASIC but any MICROSOFT language can be used. Observe the
     syntax rules for the language used.

The OPTIFUN.BAS program is

```
REM _____     PROGRAM OPTIFUN.BAS_____
END


SUB UserPenalty (wx(),wK1,wK2,wg1,wg2,wg3,wg4,wh1,wh2,wh3,wh4,wf) STATIC
' Confine your code between dashed lines only. To avoid conflict with program
' common variables, use parameter names which all start with the letter w.
' Observe the names of the parameters in the argument list. Up-to 4 equality
' constraints wg1,wg2,wg3,wg4 and 4 inequality constraints wh1,wh2,wh3,wh4
' are allowed and 10 design variables wx(1...10). Parameters wK1 and wK2 are
' defined in the interactive session later.
' Array wx() has dimension 10

      REM SUBROUTINE DEFINING PENALTY FUNCTION F[x(N)]
      REM
      REM    f[x(1),x(2),..,x(n)]=objective function +K1*(g1^2+g2^2+...)
      REM                                            +k2*[<h1>^2+<h2>^2+..]
      REM
      REM    where x(1),x(2),.... the design variables
      REM          g1,g2,...       equality constraints[functions of h(i)]
      REM          h1,h2,...       inequality constraints,        >>
      REM          <hj>=0 when hj>0, <hj>=hj if hj<0
      REM          K1,K2 weighting factors
      REM
      REM If the variables have values which are very different, the
      REM  algorithm will be slower. You can overcome this by scaling.
      REM For example, diameter=WX(1):heigth=Wx(2):thickness=WX(3)/1000
      REM Then, the program variable WX(3) has the same order of magnitude as
      REM You can do the same with the constraints to make them of the same
      REM order of magnitude with the function.
' Write your definition of the penalty function wf between dotted lines:
```

```
'-------------------------------------------------------------------
   REM ****HERE STARTS THE USER-WRITTEN CODE****
   ' Definition of problem variables
   diameter=wx(1):height=wx(2):thickness=wx(3)/1000:pi=3.14159
   '                         Container material to be minimized
   Volume = (pi*diameter*height + 2*pi*diameter^2 /4)*thickness

   '               Equality constraint (divide by 1E7 for scaling)
   wg1 = (9810*height*diameter/(2*thickness) - 10E6)/1E7
   '                         Inequality constraint:
   wh1 = 2 - pi*diameter^2/4
   IF wh1>0 THEN
        wh1Squared=wh1^2
   ELSE
        wh1Squared = 0
   END IF
   '                         Penalty function:
      wf = Volume + wK1 * wg1 ^ 2 + wK2 * wh1Squared
'Here ends the user-written code
'-------------------------------------------------------------------
END SUB
```

2. Save program as OPTIFUN.BAS.

3. Compile OPTIFUN.BAS to produce object code OPTIFUN.OBJ.

4. Link OPTIMUM.OBJ+OPTIFUN.OBJ to produce executable code OPTIMUM.EXE.

5. Run OPTIMUM.EXE through your COMPUME menu program.

7. The first window is for program identification, above. To continue, hit the ENTER key.

6. Next is then the data window.

8. You are now in the data page. In each line, the variable to be given is named, followed by a default value. To just accept the default value, simply hit ENTER. To change, type the desired value after the ? prompt and hit ENTER.

Number of variables (number of variable parameters the optimum values of which we are trying to find)

Search step length (small fraction of the expected smaller range of anyone of the variables, say range/1000)

Minimum search step length (fraction of previous step length, say search step length/10. The program starts with the search step length and when it finds an approximate optimum it reduces the step length up to the above minimum for more precise determination of the optimum point).

Minimum of the penalty function. Normally zero, when the program reaches this value it stops. Otherwise, the program stop in a prescribed number of iterations or by the user.

Acceleration factor. A value greater than 1 which multiplies the step length after each unidirectional step to accelerate the search, usually 1.5 to 2.

Deceleration factor. A value less than 1 which multiplies the step length after the approximate minimum is determined for more accuracy, usually 0.2 to 0.5

K1 Equality constraint multiplier. Start with a small value, say 10, and after finding a minimum repeat with a greater value.

K2 Inequality constraint multiplier. Start with a small value, say 10, and after finding a minimum repeat with a greater value.

Upper bound of the variables, if any. Otherwise, use very large values, say 1e20.

Lower bound of the variables, if any. Otherwise, use very small values, say -1e20.

Initial values of the variables. Good judgement here is essential. Good initial guesses help the process very much. In this example, use h(1) = 1 (diameter), h(2) = 5 (height), h(3) = 1 (thickness x 1000).

Are the data correct ? Answer N (no) if you want to change any parameter, answer Y (yes) to continue.

In the latter case, the next page, method selection, is shown:

Default values are 1 for Steepest Descent and 1 for Golden Section. Then, the execution window appears.

If there are equality and inequality constraints, their values are shown in the middle of the page. If at the program termination their value is high, the values of K1 and/or K2 must be increased.

9. Hit **M** to stop execution and return to data definition. The last found values of the variables are now the initial ones. Change the parameters desired and optimize again.

The output screen is shown below. The solution is, Diameter = 0.58 m, Height = 4.91 m, thickness = 1.40 mm.

```
POINT  2    CURRENT STEP- .01  FUNCTION-        3.321292E-03

Initial function- 17
Initial Guess Vector  Current values of constraints     FINAL OPTIMUM VECTOR:
     1      1                                                1    1.998697
     2      1                                                2    3.240033




Final Penalty Function        3.321292E-03


Optimization completed in   2 directions    Continue? (Y/N)?
                                             Type M to return to Menu
_____Optimization program OPTIMUM_____
Direction Method:Steepest Descent       Line Search Mode:Golden Section
STATUS:  Golden section search...

POINT  2    CURRENT STEP- .01  FUNCTION-        3.321292E-03

Initial function- 17
Initial Guess Vector  Current values of constraints     FINAL OPTIMUM VECTOR:
     1      1                                                1    1.998697
     2      1                                                2    3.240033
```

Probable Errors: The penalty function changes very little: small step. The penalty function changes erratically: large step. When return to data, some have different values (except for the initial values): You used a program variable name in your penalty function definition subroutine. Division by zero: Penalty function is zero due to wrong definition. Try also different initial values.

11

CHAPTER ELEVEN
FASTENERS

11.1. RIVETS

Function: Analysis of eccentric riveting.

Reference: Dimarogonas, A.D. 1988, chapter 8, Shigley.

Files needed: RIVETS.EXE, COMPUME.SYS.

Hardware Requirements: 512k, 1 FD 360k, any card, monochrome monitor.

Limitations: Up to 10 rivets.

Method: Coordinate system (x,y,z) with origin at the area center of riveting and the z-axis is perpendicular to its plane.

$$\tau_{it} = M_z r_i / (A_r \Sigma r_i^2)$$

$$\tau_{ix} = F_x/A_r + \tau_{it} x/r, \quad \tau_{iy} = F_y/A_r + \tau_{it} y/r$$

$$\sigma_{iz} = F_z/A_r + M_{xyi}/(A_r \Sigma y_i^2) + M_{yxi}/(A_r \Sigma x_i^2)$$

$$\sigma_{ieq} = \sqrt{\sigma_{iz}^2 + 4(\tau_{ix}^2 + \tau_{iy}^2)}, \text{ maximum shear stress theory.}$$

Design equation: $\sigma_{eq-max} = S_y/N$, solves for $d^2 = 4A_r/\pi$.

Use: To use the compiled program, select from MENUMEC or type RIVETS.

The first window is for program identification and basic data: Allowable shear stress, forces and moment about the riveting area center. Hit the ENTER key.

Next is the main menu:

Quit quits the program execution.

New starts a new riveting. Any previous one is erased.

Save File saves the current riveting on disc. At prompt, enter file name desired (preferable RIVET*.DAT). It prints a list of the files RIVET*.DAT for reference. If no name is entered, you return to menu.

Load File loads a frame from disk. See above.

Clear clears the screen without erasing the riveting from memory.

Data Prep: Interactive preparation of geometric data, the interactive graphics input-output window. You specify the x,y coordinates of each rivet with the cursor and the keypad arrows.

Solve Starting with a guess of the rivet diameter the program yields the rivet diameter, the maximum shear stress and the area center of the riveting.

Example: An example of a riveting is built-in into the program. Select input file RIVET1.DAT. Hit ENTER repeatedly and the example will be executed.

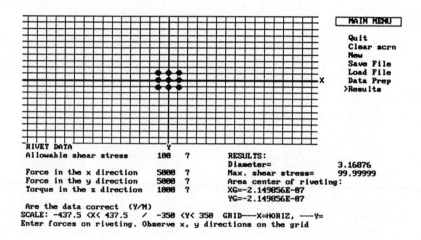

11.2. WELDS

Function: Analysis of complex welds

Reference: Dimarogonas, A.D. 1988, chapter 8.

Files needed: WELDS.EXE, COMPUME.EXE.

Hardware Requirements: 512k, 1 FD 360k, EGA card, monochrome monitor.

Limitations: Up to 10 linear and 10 circular segments. Limits can be extended by appropriate changes in the program.

Method: A coordinate system (x,y,z) is used with origin at the area center of the weld and the z-axis is perpendicular to its plane. The rivet stresses are

$$\tau_{tP} = M_z r_P / J_{wz}$$

$$\tau_{xP} = F_x / A_w + \tau_{tP} x / r, \quad \tau_{yP} = F_y / A_w + \tau_{tP} y_P / r$$

$$\sigma_{zP} = F_z / A_w + M_y y_P / J_{yy} + M_x x_P / J_{xx}$$

$$\sigma_{Peq} = \sqrt{\sigma_{zP}^2 + 4(\tau_{xP}^2 + \tau_{yP}^2)}, \text{ maximum shear stress theory.}$$

where F and M designate forces and moments, respectively.

The program tests the weld and yields $\sigma_{Peq\text{-}max}$.

Use:

To use the compiled program, select from MENUMEC or type WELDS.

The only window is for program identification and basic data, maximum x and y in the units to be used in the weld dimensions, or the maximum horizontal and vertical coordinates expected. Enter xmax, ymax, number of linear and circular segments.

For each line segment of the weld, the program asks for the coordinates of the two ends and the weld width. For each circular segment, the program asks for the center coordinates, the beginning and end angles, the arc radius and the weld width. Then the program asks for the loads on the weld, 3 forces and 3 moments. The weld plane is x,y.

The program yields the magnitude and location of the maximum stress and plots the weld in the same window.

Example: An example of a welding is built-in into the program. Select input file WELD1. Hit ENTER repeatedly and the example will be executed.

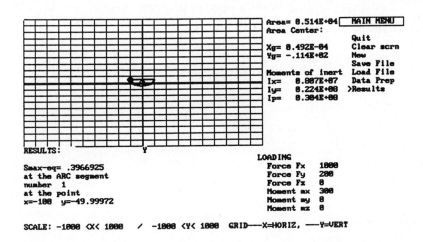

11.3. Column design for **BUCKLING**

Function: Designing a column to resist buckling.

Reference: Dimarogonas, A.D. 1988, chapter 9.

Files needed: BUCKLING.EXE

Hardware Requirements: 512k, 1 FD 360k, EGA card, monochrome monitor.

Limitations: None.

Method: The Euler and Johnson equations are used:

Euler $P_{cr} = \pi^2 EI/(\mu L)^2$ for $(L/r)^2 > \pi^2(2E/S_y)$

Johnson $P_{cr} = S_y[1-S_y(L/r)^2/(4\pi^2 E)]$, $r^2 = I/A$

$$\text{for } (L/r)^2 < \pi^2(2E/S_y)$$

P_{cr} is the critical load, E the Young's modulus, I the section moment of inertia, A the section area, ^{L} the length, S_y the yield strength. $\mu = 1$ for simply supported beam, 2 for cantilever column, 1/2 for both ends fixed.

The program solves for the column diameter, if circular column, or the minor moment of inertia, if the column is non-circular.

Use: To use the compiled program, select from MENUMEC or type BUCKLING.

The first window is for program identification and basic data, axial load, column length, material yield strength and Young modulus. Enter data at the ? prompt and hit ENTER key.

Next is then the results window.

Enter the boundary conditions from the given menu. Then, the program prints the diameter, compressive stress and the equivalent minimum moment of inertia if the section is not circular.

Example: An example of a column design is built-in into the program. Select input file COLUMN1. Hit ENTER repeatedly and the example will be executed.

```
ENTER DATA:

Enter axial load                    10000    ?
Enter column length                 1000     ?



Boundary conditions    m
_____

pinned-pinned          1
clamped-free           2
  >>   -clamped       .5
  >>   -pinned        .7

Enter m:  1      ?


SOLUTION:
_____
Diameter-........................ 12.52025
Compressive stress-.................. 81.224
Equivalent minimum moment of inertia- 1206.207

SLENDER COLUMN

Enter C to continue, hit ENTER to return to the root menu?
```

11.4. **HELIC**al **SP**rings

Function: Design of helical springs.

Files needed: HELICSP.EXE

Reference: Dimarogonas, A.D. 1988, chapter 9.

Hardware Requirements: 512k, 1 FD 360k, EGA card, monochrome monitor.

Limitations: None.

Method: The design equations are:

Flexibility: $f_1(d,D) = Gd^4/(8N_aD^3) - k = 0$

Strength: $f_2(d,D) = (8\pi D/\pi d^3)(D/d - 0.5)/(D/d - 1) - S_{sy}/N = 0$

where d the wire diameter, D the helix diameter, N_a the number of turns, k the spring constant, G the shear modulus, S_{sy}/N the allowable shear stress. The Newton-Raphson iteration is used to solve the system of equations for d, D.

  Use: To use the compiled program, select from MENUMEC or type HELICSP.

  The first window is for program identification. Hit the ENTER key.

  You are now in the data window. Enter data at the ? prompt. The program then prints the results. A parameter study menu follows.

  Select D for a new design. The program returns to the data definition.

  Select P for parameter study. The program asks for the type of parameter study: number of turns and strength. You select either one or both. The program asks then for the upper and lower limits and step for each parameter study. Finally, it prints the parameter combinations which wield feasible design and the spring volume for each.

  Example: An example of a helical spring design is built-in into the program. Select input file HELISP.DAT Hit ENTER repeatedly and the example will be executed.

  Probable Errors:

No convergence: Design not feasible. Gross error in units.

```
                    ENTER DATA
          ----------------------
          Design data:
          Spring constant      50          ?
          Maximum load         1000        ?
          Number of turns      12          ?

          Material properties:
          Shear modulus        80000       ?
          Young modulus        210000      ?
          Poisson ratio        .3          ?
          Allowable shear stress 400       ?
          Are the data correct (Y/N)?
                    RESULTS
          --------------------
          Coil Diameter  :   26.40187
          Wire Diameter  :   5.764529
          Max. Deflection:   20
          Free length    :   100.7034
          Min free length:   80.7034
          Helix angle    :   5.124814  deg
          Buckling load  :   3894.539
          HIT RETURN TO CONTINUE ?
```

12

CHAPTER TWELVE
ROTARY DRIVES

12.1. Design of **FLAT BELT**s

Function: Design of transmission flatbelts.

Reference: Dimarogonas, A.D. 1988, chapter 10.

Files needed: FLATBELT.EXE

Hardware Requirements: 512k, 1 FD 360k, EGA card, monochrome monitor.

Limitations: Available materials in program.

Method: The following belt properties are used

type	v_{max}	Ec	Sc	Sn	m	d_1/t	b_{min}	b_{max}	f
"Rubber"	40	10	2.5	6	5	30	20	300	.5
"Leather"	40	30	2.9	6	6	35	20	600	.3
"Fabric"	22.5	15	2.1	3	5	25	0	1000	.3
"Plastic"	60	60	6.1	6	6	100	10	750	.75
"Steel"	12E5	21000	330	165	6	1000	20	250	.75

where v_{max} the maximum linear velocity, E the Young's modulus, E_c the design value of the Young's modulu S_n the fatigue strength, S_c the design fatigue strength, m is the fatigue life exponent, t the thickness, d_1/t the minimum diameter of the driving pulley over the length ratio, b the width, f the friction coefficient. The program computes the maximum fatigue life for each of the above belt types, thickness and width.

First, it estimates the driving pulley diameter, from optimization considerations, as

$$d_1 = 1200(P/n_1)^{.333} \text{ mm,}$$

The design equations are

$$d_2 = 0.985 d_1 n_1 / n_2 \quad \text{mm,}$$

$$L = (2a\pi/2)(d_1 + d_2) + (d_2 - d_1)^2/4a, \text{ belt length,}$$

$$\phi = \pi - (d_2 - d_1) / 3\pi a, \text{ friction arc,}$$

α = .000011 kg/mm³,

g = 9810 mm/sec²,

p = $e^{f\phi}$,

V_{max} = $n_1(2\pi/60)d_1/2$,

N_o = $(S_c/N - V_{max}^2\alpha/g - Et/ d_1)tV_{max}(p - 1)/10^6 p$,

σ_v = $V_{max}^2\alpha/g$, σ_e = Et/d_1,

b = PC/N_o,

U = P/V_{max},

F_0 = $U(p + 1)/2(p - 1)$,

F_1 = $F_0 + U/2$, $F_2 = F_0 - U/2$,

σ_{max} = $\sigma_v + \sigma_e + F_1/bt$,

Life, Hours = $10^6 n_1(S_n/\sigma_{max})^m/(3600zV_{max}L)$,

Pull ratio = $U/2F_0$,

where from the input data, n is the speed of rotation, subscripts 1 and 2 refer to the driving and driven pulleys, respectively, C is the service factor, P the transmitted power, z the number of belt bends per full cycle of rotation, a the belt distance.

The output is, d the pulley diameters, b the belt width, t the belt thickness, F_1 and F_2 the belt tensions, F_0 the static belt tension.

```
INPUT DATA:               (Default value:)

Power transmitted (kW)       5                          ?
Speed of driving shaft (rpm)  2000                      ?
Speed of driven shaft (rpm)   1000                      ?
Distance between shafts (mm)   500                      ?
Belt thickness:  Rubber 2-10,Leather 3-20
 Fabric 2-12,Plastic 0.5-6, Steel 0.1-1.1:
Belt thicknesss       (mm)    5                          ?
Overload factor (1.0 to 1.5)
    Impact Loads (1.5 to 2.5)  1.3                      ?
Environment factor (1.0 to 1.3) 1.2                     ?
Continuous operation factor:
 Hours per day:    factor:
    3 - 4           1.45
    8 - 10          1.5
   16 - 18          1.9
     24             2.         1.9                       ?
Tension factor:
  with tension bolts : 1
  with shortening    : 1.2
  with self tension  : .8       1                        ?
Are the data correct? (Y/N)               ?


               FEASIBLE DESIGNS:
```

Material of belt	Width	Thickness	d1	d2	Length	Hours of operation	Pull factor
Rubber	105.8	5.0	163.2	321.5	1773.	116.E+006	0.606
Leather	113.1	5.0	175.0	344.8	1830.	344.E+005	0.395
Fabric	171.8	5.0	163.2	321.5	1773.	455.E+003	0.398
Plastic	23.5	5.0	500.0	985.0	3450.	324.E-003	0.662

```
Steel: Not feasible,   velocity (m/sec)= 523.5983  >  120
```

Use: To use the program (User types what below is underlined):

Select from MENUMEC or type FLATBELT.

The first window is for program identification. Hit the ENTER key.

Next is then the data window. The input data are entered at the prompt. To return to the data definition, answer N to the question "Are the data correct?"

The next window has the results for different materials.

The program asks if change in data is desired. The computation is then repeated.

Example: An example of a flat belt design is built-in into the program. Select input file FLATB.DAT. Hit ENTER repeatedly and the example will be executed.

Probable Errors:

No feasible design: too high load or too low speed for the materials available.

12.2. Design of **V-BELTS**

Function: Design of transmission V-belts.

Files needed: VBELTS.EXE

Reference: Dimarogonas, A.D. 1988, chapter 10.

Hardware Requirements: 512k, 1 FD 360k, any card, monochrome monitor.

Limitations: Available materials in program.

Method: Standard pulley diameters, mm, are:

20,22,25,28,32,36,40,45,50,56,63,71,80,90,100,112,125,140,160,180,200,224,2
50,280,315,355,400,450560,630,710,800,900,1000,1120,1250,1400,1600,
1800,2000,2240,2500,2800,3150,3550,4000,4500,5000

The standard width and thickness and the corresponding minimum driving pulley diameters are (mm):

b	t	d_{1-min}
5	3	40
6	4	50
8	5	63
10	6	80
13	8	100
17	11	132
20	12.5	180
25	16	236
32	20	315
40	25	450
50	32	600

The standard lengths are (mm)

860,1262,1916,2820,4275,6332,9540,14050,18063,18080,18100

For V-belts,

Fatigue life: Sn = 9 MPa

Stress exponent: m = 8
Friction coefficient: f = .3, apparent f_c = 0.9 due to the wedge effect.

Modulus of Elasticity: E Applicable fatigue strength S_c (MPa)

$800 < d$ (mm) , E = 44, S_c = 3.2
$500 < d < 800$, E = 35, S_c = 3.2
$315 < d < 500$, E = 28, S_c = 3.2
$200 < d < 315$, E = 21, S_c = 3
$125 < d < 200$, E = 18, Sc = 2.8
 $90 < d < 125$, E = 10, Sc = 2.5
      $d < 90$, E = 10, S_c = 2.3

The minimum diameter is computed, from optimization considerations, as

$$d_1 = 1200(P/n_1)^{0.333} \text{ mm.}$$

The design equations are

$$d_2 = 0.985 d_1 n_1 / n_2 \quad \text{mm,}$$

$$L = (2a\pi/2)(d_1 + d_2) + (d_2 - d_1)^2/4a, \text{ belt length,}$$

$$\phi = \pi - (d_2 - d_1) / 3\pi a, \text{ friction arc,}$$

$$\alpha = .000011 \text{ kg/mm}^3,$$

$$g = 9810 \text{ mm/sec}^2,$$

$$p = e^{f\phi},$$

$$V_{max} = n_1(2\pi/60)d_1/2,$$

$$N_o = (S_c - V_{max}^2 \alpha/g - Et/ d_1)btV_{max}(p - 1)/10^6 p,$$

$$\sigma_v = V_{max}^2 \alpha/g, \quad \sigma_e = Et/d_1,$$

$$n_b = Pc/N_o,$$

$$U = P/V_{max},$$

$$F_0 = U(p + 1)/2(p - 1),$$

$$F_1 = F_0 + U/2,$$

$$F_2 = F_0 - U/2,$$

$$\sigma_{max} = \sigma_v + \sigma_e + F_1/bt,$$

$$\text{Life, Hours} = 10^6 n_1(S_n/ \sigma B_{max})^m/(3600zV_{max}L),$$

$$\text{Pull ratio} = U/2F_0,$$

where from the input data, n is the speed of rotation, subscripts 1 and 2 refer to the driving and driven pulleys, respectively, C is the service factor, P the transmitted power, z the number of belt bends per full cycle of rotation, a the belt distance.

The output is, d the pulley diameters, n_b the number of belts, b the belt width, t the belt thickness, F_1 and F_2 the belt tensions, F_0 the static belt tension.

Use: To use the program (User types what below is underlined):

Select from MENUMEC or type VBELTS.

The first window is for program identification. Hit the ENTER key.

Next is then the data window.

The input data are entered at the prompt. To return to the data definition, answer N to the question "Are the data correct?" at the bottom of the page.

The next window has the results. The program yields the number of belts needed for all sections.

Example: An example of a V-belt design is built-in into the program. Select input file VBELT.DAT Hit ENTER repeatedly and the example will be executed.

```
                        V - BELT DESIGN

        INPUT DATA:                    (Default value:)

        Power transmitted (kW)         5                    ?
        Speed of driving shaft (rpm)   2000                 ?
        Speed of driven shaft (rpm)    1000                 ?
        Distance between shafts (mm)   500                  ?
        Overload factor (1.0 to 1.5)   1.3                  ?
        Envirnment factor (1.0 to 1.3) 1.2                  ?
        Continuous operation factor:
          Hours per day      factor

            3 - 4            1.45
            8 - 10           1.5
           16 - 18           1.9
              24             2.          1.9               ?
        Tension factor:
          with tension bolts : 1
          with self tension  : .8        1                 ?

        Are the data correct? (Y/N)                ?


                        FEASIBLE DESIGNS:

        Number  Width  Thickness   d1      d2    Length    Hours of    Pull
        of belts                                           operation   factor


          42     5.        3.      180.    355.   1856.    329.E+005   0.848
          27     6.        4.      180.    355.   1856.    155.E+005   0.848
          17     8.        5.      180.    355.   1856.    782.E+004   0.848
          12    10.        6.      180.    355.   1856.    416.E+004   0.848
           7    13.        8.      180.    355.   1856.    602.E+003   0.848
           5    17.       11.      180.    355.   1856.    135.E+002   0.848
           4    20.       13.      180.    355.   1856.    627.E+001   0.848
           3    25.       16.      250.    500.   2209.    521.E+000   0.827
           5    32.       20.      315.    630.   2534.    219.E-001   0.806
        Do you want to change data (Y/N) ?
```

Probable Errors: No feasible design: too high load or too low speed for the materials available. Very great number of belts: data error, V-belt is no feasible design.

12.3. GEAR PLOT

Function: Plotting of involute gear teeth.

Files needed: GEARPLOT.EXE, COMPUME.SYS

Reference: Dimarogonas, A.D. 1988, chapter 12.

Hardware Requirements: 512k, 1 FD 360k, EGA card, monochrome monitor.

Limitations: None.

Method: The tooth flanks have involute geometry outside the basic circle and radial direction otherwise. The addendum height is m and the dedendum height is 1.25 m, m is the module. The tooth roots are not rounded.

Use: To use the program (User types what below is underlined):

Select from MENUMEC or type <u>GEARPLOT.</u>

The first window is for program identification and basic data. The unit of length is mm. Enter the modul and the number of teeth of the gears. Enter a scale factor. With proper selection of the scale factor, one can plot both full gears (small scale factor) or the detail about the point of contact (big scale). The angle of rotation of the gears is in respect to the position when the contact is on the centerline. Hit the ENTER key to plot the gears.

Next is then the results window, in which the gears are shown in animated motion. Increase the speed of rotation by hitting f, decrease it by hitting s.

Example: An example of spur gear plot is built-in into the program. Select input file GEARPLOT.DAT. Hit ENTER repeatedly and the example will be executed.

Probable Errors: Teeth overlap: small number of teeth

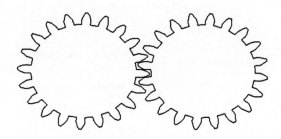

Hit E to exit, F for faster, S for slower

12.4. GEAR DESign

Function: Design of spur, helical and bevel gears.

Available Code: .EXE

Files needed: GEARDES.EXE

Reference: Dimarogonas, A.D. 1988, chapter 12.

Hardware Requirements: 512k, 1 FD 360k, EGA card, monochrome monitor.

Limitations: None.

Method: The AGMA method is used

Use: To use the program (User types what below is underlined):

Select from MENUMEC or type GEARDES.

The first window is for program identification. Hit the ENTER key.

Next is then the main menu and data window.

Enter first the type of gearing. Then, enter the data after each ? prompt. To retain the default values, simply hit ENTER. At the end, the program asks if the data are correct. If the answer is N, the program returns to the data entry. Otherwise, it proceeds with the design and results.

The program allows for redesign, changing the basic parameters. One can do that or return to the data definition for data change. If a run is repeated with the same data and the results are slightly different, this is due to the fact that the program iterates a limited number of times each run.

Example: An example of spur gear design is built-in into the program. Select input file GEARDES.DAT. Hit ENTER repeatedly and the example will be executed.

Probable Errors:

The program does not converge: Extreme values of data, unit errors.

No solution: Design not feasible. Change design parameters.

```
            GEAR DESIGN PROGRAM
    Problem identification:GEAR DESIGN PROBLEM    ?
            SPUR GEARS         1  Your selection  1 ?
            HELICAL GEARS      2      Spur gears
            BEVEL GEARS        3
DATA (Units N, mm, sec):...hit ENTER for default value
Driving power , kw        16 ?    Pinion speed,RPM 3500 ?
Gear   ratio               4 ?    Pinion No of Teeth 20 ?
No of cycles          100000 ?    Asymmetry factor  1.4 ?
SURFACE BHN: Pinion      220 ?    Gear              220 ?
MODULUS E: Pinion     210000 ?    Gear           210000 ?
FATIGUE STRENGTH:Pinion  170 ?    Gear              170 ?
SURFACE STRENGTH:Pinion  680 ?    Gear              680 ?
Service Factor Ko       1.25 ?    Width/a ratio      .3 ?
Reliability %             90 ?    Temperature,deg C  60 ?
Involute angle (14.5,20,25)              20 ?
Precision: High <1>,Ground <2>,Non-ground <3>  2 ?

Are the data correct (Y/N) ?
```

```
RESULTS:   GEAR DESIGN PROBLEM
                         PINION        GEAR
-------------------------------------------------
No of teeth                22           88
Diameter                   44          176
BHN                       220          220
Young modulus          210000       210000
Fatigue strength          170          170
Surface strength          680          680

SYSTEM PARAMETERS:
Center distance    110          Width           33
Modul              2            b/a ratio       .3
Pressure angle     20           Tangential force 1984.416
Diametral pitch    12.7         Radial force    721.8706
Contact ratio      1.607513     Axial force     0
                                Gearing volume/1E6 .8525854

DO YOU WANT: <1> Changes, <2> New Design,<3> QUIT ?
```

12.5. WORM DESign

Function: Design of worm-gears.

Files needed: WORMDES.EXE

Reference: Dimarogonas, A.D. 1988, chapter 12.

Hardware Requirements: 512k, 1 FD 360k, EGA card, monochrome monitor.

Limitations: None.

Method: The AGMA method is used.

Use: To use the program (User types what below is underlined):

Select from MENUMEC or type <u>WORMDES</u>.

The first window is for program identification. Hit the ENTER key.

Next is then the main menu and data window.

Enter the data after each ? prompt. To retain the default values, simply hit ENTER. At the end, the program asks if the answer is N, the program returns to the data entry. Otherwise, it proceeds with the design and results.

```
                          WORM DESIGN PROGRAM
                   Problem Identification:Test Example?


        DATA (Units N, mm, sec):   hit ENTER for default value

        Driving power , kw        30 ?      Worm speed, RPM 1000 ?
        Gear   ratio              15?       Worm No of leads   2  ?
        No of cycles          100000 ?      Asymmetry factor 1.4 ?
        SURFACE BHN:      Worm    200 ?         Gear          200 ?
        MODULUS E:        Worm 210000 ?         Gear       210000 ?
        FATIGUE STRENGTH:Worm     150 ?         Gear           30 ?
        SURFACE STRENGTH:Worm     600 ?         Gear          150 ?
        Service Factor    Ko     1.25 ?     Width/a ratio      .2 ?
        Reliability %             90  ?     Temperature,deg C 100?
        Involute angle (14.5,20,25)           20 ?
        Precision: High <1>,Ground <2>,Non-ground <3>  3 ?


        Are the data correct (Y/N) ?

          RESULTS, JOB Test Example.   UNITS: N, mm, sec

                              WORM          GEAR
        --------------------------------------------------------
        No of teeth           2             30
        Diameter              45.3256       113.314
        BHN                   200           200
        Young mod             210000        210000
        Fatigue strength      150           30
        Surface strength      600           150

        SYSTEM PARAMETERS:                  Gear Width    22.6628
        Center distance    79.3198          b/a ratio        .2
```

The program allows for redesign, changing the basic parameters. One can do that or return to the data definition for data change. If a run is repeated with the same data and the results are slightly different, this is due to the fact that the program iterates a limited number of times each run.

Example: An example of worm-gear design is built-in into the program. Select input file WORMDES.DAT. Hit ENTER repeatedly and the example will be executed.

Probable Errors:

The program does not converge: Extreme values of data, units errors.

No solution: Design not feasible. Change design parameters.

12.6. **SIMU**lation of rotary **DRIV**es

Function: Dynamic simulation of a two rotor torsional system.

Reference: Dimarogonas, A.D. 1988, chapter 14.

Hardware Requirements: 512k, 1 FD 360k, EGA card, color monitor.

Limitations: None.

Method: The system is modeled as a motor rotor connected with a driven rotor via en elastic shaft. The differential equations of motion are

$$J_1\theta_1^{\circ\circ} = T(\theta_1^{\circ}) + k(\theta_2 - \theta_1) + c(\theta_2^{\circ} - \theta_1^{\circ})$$
$$J_2\theta_2^{\circ\circ} = -M(\theta_2^{\circ}) - k(\theta_2 - \theta_1) - c(\theta_2^{\circ} - \theta_1^{\circ})$$

where J are the polar moment of inertia, θ the angle of rotation, 1 is the motor, 2 is the driven machine, T is the motor torque, M is the driven machine torque, both functions of the angular velocity, k is the coupling stiffness, c is the coupling damping. Moreover, the $M(\theta^{\circ})$ function is assumed to be a cubic function (pump) $M_0(\omega/\omega_0)^3$, where M_0 and $\omega_0 = (\theta^{\circ})_0$ are the rated torque and speed. The motor is assumed to have a function $T(\theta^{\circ})$:

$$T(\theta^{\circ}) = T_0(2 - \theta^{\circ}/0.95\omega_0) \text{ for } 0 < \theta^{\circ} < 0.95\omega_0$$
$$= 20T_0(1 - \theta^{\circ}/\omega_0) \quad \text{for } 0.95\omega_0 < \theta^{\circ} < \omega_0$$
$$= T(2\omega_0 - \theta^{\circ}) \qquad \text{for } \omega_0 < \theta^{\circ}$$

The differential equations of motion are solved numerically and the speeds of the driving and driven rotor are plotted, together with the torque.

Use: To use the program (User types what below is underlined):

Select from MENUMEC or type <u>SIMUL</u>.

The first window is for program identification. Hit ENTER.

Next is the data window.

Enter data at the ? prompt and hit ENTER. The program will then transfer to the results window.

Example: An example of a motor-pump system simulation is built-in into the program. Select input file SIMUL.DAT. Hit ENTER repeatedly and the example will be executed.

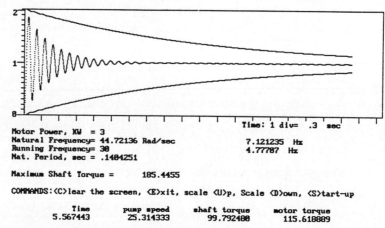

```
Motor Power, KW  = 3                              Time: 1 div=  .3  sec
Natural Frequency= 44.72136 Rad/sec                  7.121235  Hz
Running Frequency= 30                                4.77787  Hz
Nat. Period, sec = .1484251

Maximum Shaft Torque =        185.4455

COMMANDS:<C>lear the screen, <E>xit, scale <U>p, Scale <D>own, <S>tart-up

            Time          pump speed       shaft torque      motor torque
          5.567443        25.314333         99.792480         115.618889
```

Probable Errors: Overflow: Small time integration step. Change it at line 340.

Division by zero: Zero data.

12.7. Design of **SHRINKFiT** joints

Function: Design Shrink fit of a compound wheel on a shaft.

Reference: Dimarogonas, A.D. 1988, chapter 14.

Files needed: SHRINKFT.EXE, COMPUME.SYS

Hardware Requirements: 512k, 1 FD 360k, EGA card, monochrome monitor.

Limitations: 10 wheel sections.

Method: The hub is assumed to consist of a number of rings with widths b_i and inner radii r_i. If the interface state vector is $\{z\} = \{\,p\ u\,\}$, where p is the pressure and u the radial displacement, the transfer matrices are

$$[B] = \begin{bmatrix} -a_{11}/a_{12} & 1/a_{12} \\ a_{21}-a_{11}a_{22}/a_{12} & a_{22}/a_{12} \end{bmatrix}, \text{ field}, \quad [C] = \begin{bmatrix} b_i/b_{i+1} & 0 \\ 0 & 1 \end{bmatrix}, \text{ point},$$

where $a_{11} = [(1+\nu)r_i^3 + (1-\nu)r_i r_{i+1}^2]/[E(r_{i+1}^2 - r_i^2)]$,

$\qquad a_{12} = [-(1+\nu)r_i^2 r_{i+1} - (1-\nu)r_i r_{i+1}^2]/[E(r_{i+1}^2 - r_i^2)]$

From the inner to the outer diameter, the transfer matrix sequence gives

$$\{z_n^+\} = [B_n][C_n]\ldots[B_2][C_2][B_1]\{z_1^-\} = [L]\{z_1^-\}$$

Since the outer pressure is zero and the inner pressuer is p, we obtain

$$u_1 = (-L_{11}/L_{12})p_1$$

relating the inner pressure with the inner radial displacement. The pressure must be sufficient to overcome the torque

$$p = 2T/f\pi d^2 L\omega$$

The interference required is then

$$u = (-L_{11}/L_{12})2T/f\pi d^2 L\omega$$

Use: To use the program (User types what below is underlined):

Select from MENUMEC or type <u>SHRINKFT</u>.

The first window is for program identification. Hit ENTER.

Next is the data window.

```
                    ENTER DATA:
    Problem identification:   Test Shrink-fit?
    Number of rings          4 ?
    Ring No.  1                                                     RESULTS:
    inner diameter    60 ?        width            60 ?
    Young Modulus    210000 ?     Poisson ratio   .27 ?          Minimum interference      - 1.993593E-04
    Ring No.  2
    inner diameter    80 ?        width            20 ?          Maximum interference      - 9.453977E-03
    Young Modulus    210000 ?     Poisson ratio   .27 ?
    Ring No.  3                                                  Min. interference pressure - .5904456
    inner diameter   100 ?        width            30 ?
    Young Modulus    210000 ?     Poisson ratio   .27 ?          Max  interference pressure - 28
    Ring No.  4
    inner diameter   140 ?        width            50 ?          With maximum stress        - 100
    Young Modulus    210000 ?     Poisson ratio   .27 ?

    Outer diameter    60 ?                                       Enter R to return to the root menu, C to continue?
    FORCES:
    Transmitted torque       100 ?     Transmitted axial force  2000 ?
    Coefficient of friction .3 ?       Allowable stress at fit  100 ?
```

Enter data at the ? prompt and hit ENTER. The program will then transfer to the results window.

Example: An example of a compound shrink fit is built-in into the program. Select input file SHRINK.DAT. Hit ENTER repeatedly and the example will be executed.

Probable Errors:

Overflow: Very rigid sections. Wrong units. Division by zero: Zero data, stiff problem.

12.8. **ROTOR DYN**amic analysis

Function: Critical speeds of rotating shaft with the transfer matrix method.

Files needed: ROTORDYN.EXE, COMPUME.SYS

Reference: Dimarogonas, A.D., Haddad, S.D., 1992, chapter 10.

Hardware Requirements: 512k, 1 FD 360k, EGA card, color monitor.

Limitations: Up to 10 nodes.

Method:

a) Natural vibration

Consider a beam with n-1 massless elements (fields) and n nodes.

We assume a beam, say simply supported, which consists of n-1 *beam elements* of different but constant section moment of inertia. Thus the beam has n-1 elements with constant cross-section and n *nodes*, points (or planes) which define the beginning or the end of a beam element.

```
                    Number of Elements      4              ?
                    Modulus of Elasticity  2.1E+11         ?
                    Material Density        7800           ?
                Element data                      node data
   El/No
            length    diameter force    spring    mass    inertia
       1   ? 1      ? .1      ?         ?         ?        ?
       2   ? 1      ? .12     ?         ?         ?        ?
       3   ? 1      ? .13     ? 1000    ?         ? 1000   ? 30
       4   ? 1      ? .1      ?         ?         ?        ?
       right end....          ?         ?         ? 100    ? 1



    Enter no of solid supports  2                          ?
    Enter numbers of support nodes:

    Support  1   at node  1                                ?
    Support  2   at node  4                                ?
    Are data correct (Y/N)  ? █
```

To fully describe the situation at each node, we need to know 4 quantities: The deflection y, the slope θ, the moment M and the shear force V. The section j of a beam between nodes j and j+1 is shown in the above figure with the usual in statics sign conventions.

These four quantities can be arranged in a vector s=$\{y\ \theta\ M\ V\}$ which, because it describes the state of the system at node j is called *state vector* and to designate the node j we shall use it with a subscript j.

Let us suppose that at the node 1 the state vector is

$$z_1 = \{x_1\ \theta_1\ M_1\ V_1\}$$

yet unknown. If no force is acting between nodes 1 and 2, the deflection, slope, moment and shear at node 2, from simple beam theory, will be, in matrix form:

$$\underline{z}_2 = \underline{L}_1\underline{z}_1$$

where

$$[L_1] = \begin{bmatrix} 1 & L & L^2/2EI & L^3/2EI \\ 0 & 0 & 1 & L \\ L^2/2EI & 0 & 0 & 0 \\ 1 & L & 0 & 0 \\ 0 & 0 & 0 & 1 \end{bmatrix},\ z_1 = \begin{bmatrix} y \\ \theta \\ M \\ V \end{bmatrix}_1,\ z_2 = \begin{bmatrix} y \\ \theta \\ M \\ V \end{bmatrix}_2$$

and the subscript 1 of the matrix indicates that quantities L, E, I are properties of the element number 1. The upper left 4x4 part of the matrix $\underline{L}$ will be used in some analyses. The fifth column and row are added here for computational convenience.

Equation 10.31 tells us that the state vector at node 2 is the state vector at node one multiplied by a square 5x5 matrix $\underline{L}$ which depends on the element properties only and it is well known. This matrix transferred the state from node 1 to node 2 and therefore it shall be called *transfer matrix*. For every element of the beam there exists one, known, transfer matrix $\underline{L}$. We can repeat the procedure for elements 2,3,..., to obtain, using also the previous relations,

A free body diagram shows the shear forces on the left and right. For harmonic motion, the acceleration of the mass m_j will be $y^{\circ\circ}{}_j = -\omega^2 y_j$. Therefore, Newton's Law gives

$$- v_j{}^R + v_j{}^L = -\omega^2 m_j y_j$$

Therefore, we can write

$$\underline{z}_j{}^R = \underline{L}_j{}^M \underline{z}_j{}^L$$

where

$$\underline{P} = \begin{bmatrix} 1 & 0 & 0 & 0 & 0 \\ 0 & 1 & 0 & 0 & 0 \\ 0 & 0 & 1 & 0 & 0 \\ \omega^2 m & 0 & 0 & 1 & 0 \\ 0 & 0 & 0 & 0 & 1 \end{bmatrix}$$

We shall obtain

$$\underline{z}_n{}^L = \underline{L}_{n-1} \underline{P}_{n-1} \underline{L}_{n-2} \cdots \underline{P}_3 \underline{L}_2 \underline{P}_2 \underline{L}_1 \ \underline{z}_1 = \underline{A} \underline{z}_1{}^R$$

Application of the boundary conditions will yield a system of 4 homogeneous algebraic equations. The condition for existence of solution will yield, because the last equation will be the identity 1=1, a 4x4 determinant equal to zero thus the frequency equation.

For large number of masses, if a computer is used, one cannot proceed with explicit evaluation of the frequency equation. Instead, the chain multiplication of matrices will yield a relation, in explicit form,

$$\begin{bmatrix} y \\ \theta \\ M \\ v \end{bmatrix}_n = \begin{bmatrix} a_{11} & a_{12} & a_{13} & a_{14} \\ a_{21} & a_{22} & a_{23} & a_{24} \\ a_{31} & a_{32} & a_{33} & a_{34} \\ a_{41} & a_{42} & a_{43} & a_{44} \end{bmatrix} \begin{bmatrix} y \\ \theta \\ M \\ v \end{bmatrix}_1$$

We shall have four boundary conditions in addition. For example, for a simply supported beam, we have $y_1 = y_n = 0$, $M_1 = M_n = 0$ and then

$$0 = a_{12}\theta_1 + a_{14}v_1$$

$$\theta_n = a_{22}\theta_1 + a_{24}v_1$$

$$0 = a_{32}\theta_1 + a_{34}v_1$$

$$v_n = a_{42}\theta_1 + a_{44}v_1$$

or, in terms of the unknowns θ_1, θ_n, V_1, V_n:

$$a_{12}\theta_1 \qquad\qquad + a_{14}\,V_1 \qquad\qquad = 0$$
$$a_{22}\theta_1 - \theta_n + a_{24}V_1 \qquad\qquad = 0$$
$$a_{32}\theta_1 \qquad\qquad + a_{34}V_1 \qquad\qquad = 0$$
$$a_{42}\theta_1 \qquad\qquad + a_{44}V_1 - V_n \qquad = 0$$

The condition for existence of solution for this homogeneous system of linear algebraic equations is

$$D(\omega) = \begin{vmatrix} a_{12} & 0 & a_{14} & 0 \\ a_{22} & -1 & a_{24} & 0 \\ a_{32} & 0 & a_{34} & 0 \\ a_{42} & 0 & a_{44} & -1 \end{vmatrix} = 0$$

The values of ω satisfying this equation are the natural frequencies.

The matrix A does not depend on the boundary conditions, but only on ω and the properties of the beam. The boundary conditions show up in the structure of the frequency determinant.

A usual situation is when at the end of the beam there is a supporting spring of constant k. The boundary conditions will be this end

$$M = 0, \quad V = ky$$

Therefore, if the spring is at the left end,

$$0 = a_{12}\theta_1 + (a_{14} + 1/k)\,V_1$$
$$\theta_n = a_{22}\theta_1 + (a_{24} + 1/k)\,V_1$$
$$0 = a_{32}\theta_1 + (a_{34} + 1/k)\,V_1$$
$$V_n = a_{42}\theta_1 + (a_{44} + 1/k)\,V_1 - V_n$$

and the frequency determinant

$$D(\omega) = \begin{vmatrix} a_{12} & 0 & a_{14} + 1/k & 0 \\ a_{22} & -1 & a_{24} + 1/k & 0 \\ a_{32} & 0 & a_{34} + 1/k & 0 \\ a_{42} & 0 & a_{44} + 1/k & -1 \end{vmatrix}$$

b) Forced harmonic vibration

As in the static case, we shall assume that at the node j there is a harmonic force $F_0\cos\omega t$. Therefore, all the components of the state vectors will be of the form $y\cos\omega t$, $\theta\cos\omega t$, $M\cos\omega t$, $V\cos\omega t$. In all equations we shall divide throughout with $\cos\omega t$ and thus deal with amplitudes.

Multiplying the transfer matrices from left to right, we shall obtain

$$z_n^R = \underline{A} z_n^L$$

The last of these equations is the identity 1=1. The first four, supplemented by boundary conditions, have four unknowns. Because they are non-homogeneous (they include F explicitly) they can be solved to yield the end conditions. The state vectors at the several nodes can be then calculated as usual from the first node.

For natural vibration problems one can omit the last column and last row of the matrices, since they lead to identity and use 4x4 matrices for lateral vibration, 2x2 for torsional vibration.

Use:

To use the program (User types what below is underlined):

Select from MENUMEC or type <u>ROTORDYN</u>. The main menu appears. You can load, save, make a data file or compute. The latter command opens the analysis menu. You can select <critical speeds> od <dynamic response>.

A file can be made with the <Make model> command:

Number of elements: The maximum expected number for dimensioning purposes.

Modulus of elasticity: Use a consistent system of units. SI is suggested.

Density: Enter material density.

Every line corresponds to a node or element numbered consecutively from left to right. Every time you hit ENTER, the cursor goes to the next entry. To enter a new value, type it and hit ENTER. For zero or to maintain the default or the previously defined value, just hit ENTER. The first 3 entries are for the element:

Length of the element, distance of node from previous one.

Diameter, for circular section, area moment of inertia for a general section.

The next 2 entries in the line are for node data:

Lateral harmonic force on the node

Spring constant of flexible support, if such support exists. Do not include here solid supports.

If you enter zero length or section size, the line is repeated.

The last line has node data only because there is one node more than elements.

Then the program asks for the number of solid supports. Enter their number and node numbers of their location, as the program requests them.

Finally, the program asks if data are correct. If the answer is no, the program returns in the beginning. The data entered already are printed on their position. Type only the corrections or additions. For all correct entries, just hit ENTER. You can repeat this as many times as you like.

If the data are correct, the program returns to the menu. You should save the data into a file, if you wish. Then, you proceed with the analysis. The program prints the deflection, slope, moment, shear and deflection at each node. Note the finite deflections at the supports. This is due to the way supports are modelled, by the program, as very hard springs. They have to be 2-3 orders of magnitude smaller than the other deflections. Otherwise, there is a numerical accuracy problem.

Example: An example of the static loading of a cylindrical shaft is built-in into the program. Select circular cross-section from the first page and input file ROTOR1.DAT Hit ENTER repeatedly and the example will be executed. The input file, which can also be made in a text editor, is:

```
PROPERTIES, 4 , 2.1E+11 , 7800
ELEMENT, 1 , 1 , .1
ELEMENT, 2 , 1 , .12
ELEMENT, 3 , 1 , .13
ELEMENT, 4 , 1 , .1
NODE, 1 , 0 , 0 , 0 , 0
NODE, 2 , 0 , 0 , 0 , 0
NODE, 3 , 1000 , 0 , 1000 , 30
NODE, 4 , 0 , 0 , 0 , 0
NODE, 5 , 0 , 0 , 100 , 1
SUPPORT, 1
SUPPORT, 4
FILEND
```

The results are plotted on the screen in the form of displacement together with the beam, the supports and the loading. The results screen for ROTOR1.DAT is shown below:

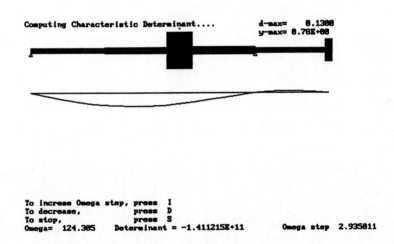

Probable Errors: Execution breaks: Hardware incompatibility, wrong graphics adaptor. Out of memory. Deflections at supports large: Numerical problems, large D/L ratio for some element. Try changing units.

12.9. **SHAFT** Design

Function: Shaft design and antifriction bearing selection.

Files needed: SHAFTDES.EXE, COMPUME.EXE, DEEPGR.AFB, ROLLER.AFB, SELFBALL.AFB, TAPER.AFB

Reference: Dimarogonas, A.D., Computer Aided Machine Design, Prentice-Hall 1988, chapter 14.

Hardware Requirements: 512k, 1 FD 360k, EGA card, monochrome monitor.

Limitations: Up to 50 nodes, 50 elements, 10 solid supports.

Method: We assume a shaft which consists of n-1 *beam elements* of different but constant section moment of inertia.

Thus the shaft has n-1 elements with constant cross-section and n *nodes*, points (or planes) which define the beginning or the end of a beam element.

To fully describe the situation at each node, we need to know 4 quantities: The deflection y, the slope θ, the moment M and the shear force V. The section j of a beam between nodes j and j+l with the usual in statics sign conventions.

These four quantities can be arranged in a vector s={y θ M V} which, because it describes the state of the system at node j is called *state vector* and to designate the node j we shall use it with a subscript j.

Let us suppose that at the node 1 the state vector is

$$z_1 = \{x_1 \ \theta_1 \ M_1 \ V_1\}$$

yet unknown. If no force is acting between nodes 1 and 2, the deflection, slope, moment and shear at node 2, from simple beam theory,

$$\underline{z}_2 = \underline{L}_1 \underline{z}_1$$

where

$$\underline{L}_1 = \begin{bmatrix} 1 & L & L^2/2EI & L^3/6EI & 0 \\ 0 & 1 & L/EI & L^2/2EI & 0 \\ 0 & 0 & 1 & L & 0 \\ 0 & 0 & 0 & 1 & 0 \\ 0 & 0 & 0 & 0 & 1 \end{bmatrix}, \quad \underline{z}_1 = \begin{bmatrix} y \\ \theta \\ M \\ V \\ 1 \end{bmatrix}_1, \quad \underline{z}_2 = \begin{bmatrix} y \\ \theta \\ M \\ V \\ 1 \end{bmatrix}_{\substack{1 \\ 2}}$$

and the subscript 1 of the matrix indicates that quantities L, E, I are properties of the element number 1. The upper left 4x4 part of the matrix $\underline{L}$ will be used in some analyses. The fifth column and row are added here for computational convenience. We can repeat the procedure for elements 2,3,..., to obtain, using also the previous relations,

$$\underline{z}_2 = \underline{L}_1 \ \underline{z}_1$$
$$\underline{z}_3 = \underline{L}_2 \ \underline{z}_2 = \underline{L}_2 \ \underline{L}_1 \ \underline{z}_1$$
$$\underline{z}_4 = \underline{L}_3 \ \underline{z}_3 = \underline{L}_3 \ \underline{L}_2 \ \underline{L}_1 \ \underline{z}_1$$

- - - - - - - - - - - - -

At the nodes, the state vector as we approach the node from left and right is the same. However, if at the node we have a static force F, this is not true. For a small length about the node, the deflection, slope and moment remain unchanged but, in order to have equilibrium, we must have $V^R = V^L + F$, where with superscript L we designate the situation at the left of the node and R refers to the situation at the right of the node.

We can write

$$y^R = y^L$$
$$\theta^R = M^L$$
$$M^R = M^L$$
$$V^R = V^L + F$$

We can write this in matrix form as

$$\underline{z}_j^{\,R} = \underline{P}\underline{z}_j^{\,L}$$

where

$$\underline{P}_2 = \begin{bmatrix} 1 & 0 & 0 & 0 & 0 \\ 0 & 1 & 0 & 0 & 0 \\ 0 & 0 & 1 & 0 & F \\ 0 & 0 & 0 & 1 & 0 \\ 0 & 0 & 0 & 0 & 1 \end{bmatrix}_2$$

$\underline{L}$ is called *field* matrix (and the element i,j called *field*), and $\underline{P}$ *point matrix*. Therefore, for transferring from left to right of a loaded node, we have to multiply with the point matrix. Then, the last of the equations will be:

$$\underline{z}_n^{\,L} = \underline{L}_{n-1}\underline{L}_{n-2} \cdots \underline{L}_j \underline{P}_j \underline{L}_{j-1} \cdots \underline{L}_2 \, \underline{L}_1 \, \underline{z}_1^{\,R}$$

In this product one can take into account any number of loads at the nodes by multiplying with all the point matrices.

Let $\underline{A} = \underline{P}_n\underline{L}_{n-1}\underline{P}_{n-1}\underline{L}_{n-2} \cdots \underline{P}_3\underline{L}_2\underline{P}_2\underline{L}_1\underline{P}_1$, a square 5x5 matrix which can be computed easily by multiplication of a chain of 5x5 matrices. Then,

$$\underline{z}_n^{\,L} = \underline{A}\underline{z}_1^{\,R}$$

The first four of this matrix equation can be written as

$$Y_n = a_{11}y_1 + a_{12}\theta_1 + a_{13}M_1 + a_{14}V_1 + b_{14}\,F$$
$$\theta_n = a_{21}y_1 + a_{22}\theta_1 + a_{23}M_1 + a_{24}V_1 + b_{24}\,F$$
$$M_n = a_{31}y_1 + a_{32}\theta_1 + a_{33}M_1 + a_{34}V_1 + b_{34}\,F$$
$$V_n = a_{41}y_1 + a_{42}\theta_1 + a_{43}M_1 + a_{44}V_1 + b_{44}\,F$$

We have four equations with 8 unknowns, the end conditions y_1, θ_1, M_1, y_n, θ_n, V_1, M_n, V_n. However, because of the boundary conditions we know four of these quantities. For example, for a simply supported beam we shall have $y_1 = y_n = 0$ and $M_1 = M_n = 0$. Therefore, we have four unknowns θ_1, θ_n, V_1, V_n 4 non-homogeneous equations to compute them.

Use: To use the program (User types what below is underlined):

Select from MENUMEC or type <u>SHAFTDES</u>. The first window is for program identification. Hit ENTER. The program transfers to the main menu: The selections are made in the order they appear. For example, you cannot plot the rotor before it is designed.

Selection of MAKE MODEL invokes the data window. Enter at the ? prompt:

Problem Identification: Any alphanumeric text.

Number of elements: The maximum expected number for dimensioning purposes.

Next, enter node and element data. Using the keypad direction keys, move around and fill only the non-zero data. You leave the node and element data by moving the cursor vertically below them. Then the program asks for material data and the service factor. Finally, the program asks if the dead weight of the shaft in the vertical direction is to be taken into account (**The load data are only external loads and NOT the dead weight and the reactions**).

Finally, the program asks if data are correct. If the answer is no, the program returns in the beginning. The data entered are printed on their position. Type only the corrections or additions. You can repeat this as many times as you like.

If the data are correct, the program returns to menu.

Design Shaft, proceeds with the analysis. It prints the computed shaft diameters.

Select bearings, proceeds to the selection of antifriction bearings. Make sure that the data files with .BAL and .ROL and other relevant extensions are on the current drive.

First, it asks which bearing takes the thrust. Specify one only such bearing.

For each bearing:

The program requests the type of bearing.

Then, it searches the files for a bearing of the specified type at the given shaft diameter. If it finds one, it prints its characteristics. If not, it asks if you wish a different type (heavier), a greater shaft diameter or a manual selection of bearing. In the latter case, the program prints the static and dynamic capacity, allowing for catalog selection.

Plot Shaft and Forces plots the shaft and the loads on it.

Example: An example of shaft on rolling element bearings is built-in into the program. Select input file SHAFT1.DAT. Hit ENTER repeatedly and the example will be executed.

```
LABEL,no label
PROPERTIES, 3 , 2 , 7800 , 2 , 2.1E+11 , 1E+08 , 2E+08 ,y
ELEMENT, 1 , .8 , 1000 , 0 , 0 , 1000
ELEMENT, 2 , .5 , 1000 , 0 , 0 , 1000
ELEMENT, 3 , .3 , 1000 , 0 , 4 , 1000
NODE, 1 , 1000 , 0 , 0 , 0 , 0 , 0 , 0 , 0
NODE, 2 , 10000 , 0 , 0 , 0 , 0 , 0 , 0 , 0
NODE, 3 , 0 , 0 , 0 , 0 , 0 , 0 , 0 , 0
NODE, 4 , 0 , 0 , 0 , 0 , 0 , 0 , 0 , 0
SUPPORT, 1 , 1
SUPPORT, 2 , 4
FILEND
```

The last entry in the properties line is x or y, indicating that gravity in the x or y direction should be taken into account.

Probable Errors: Execution breaks: Hardware incompatibility. Deflections at supports large: Numerical problems. Very short (large D/L) elements. Try changing units.

```
     EXTERNAL Horizontal force on the node (not the reaction)  Hit X to exit

     no label             Node number or element number
                          1         2         3         4

     NODE DATA:
     Horizontal Force >1000        10000     0         0
     Vertical Force    0           0         0         0
     Horizontal Moment 0           0         0         0
     Vertical Moment   0           0         0         0
     Spring-Horizontal 0           0         0         0
     Spring-Vertical   0           0         0         0
     Rotary-spring Hor 0           0         0         0
     Rotary-spring Ver 0           0         0         0
     Support(1/0)      1           0         0         1
     ELEMENT DATA
     Length            .8          .5        .3        0
     Distrib hor load  1000        1000      1000      0
     Distrib Ver load  0           0         0         0
     Torque            0           0         4         0
     Thrust            1000        1000      1000      0
     Enter Material Data:                    Young Modulus    2.1E+11   ?
      Fatigue strength 1E+08       ?         Yield Strength   2E+08     ?
      Density          7800        ?         Service Factor   2         ?
     Do you want to include the dead weight (Y/N) y?
```

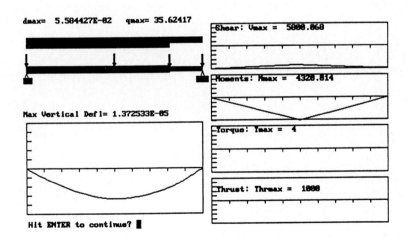

137

13

CHAPTER THIRTEEN
LUBRICATION

13.1. **SLIDER BeaRing** analysis

Function: Finite Element Analusis of hydrodynamic sleeve bearings.

Files needed: SLIDERBRG.EXE, COMPUME.SYS, one MESH*.* file.

Prerequisites: Program AUTOMESH must be used to create the triangular mesh. A file anyname.MES must exist in the current directory before you use SLIDERBR.

Reference: Dimarogonas, A.D. 1988, chapter 11.

Hardware Requirements: 512k, 1 FD 360k, EGA card, color monitor.

Limitations: Up to 100 nodes, 100 elements.

Method: Solution of the Reynolds equation

$$\frac{\partial}{\partial x}\left[\frac{h^3}{12}\frac{\partial p}{\partial x}\right] + \frac{\partial}{\partial z}\left[\frac{h^3}{12}\frac{\partial p}{\partial z}\right] = \frac{U}{2}\frac{\partial h}{\partial x} + \overset{.}{h}$$

with the finite element method.

Use:

To use the program (User types what below is underlined):

Select from MENUMEC or type FINLUB.

The first window is for program identification. Hit ENTER.

Next is then the data window.

Enter data at the ? prompt and hit ENTER. Hit ENTER only to accept default data. If data are correct, answer Y to the appropriate question.

Next is then the main menu window.

You are now in the main menu. You select with the keypad direction keys. The > sign indicates the selection and at the bottom line there is an explanation of the command. ENTER invokes the respective command.

You are now in the main menu. You select with the keypad direction keys. The > sign indicates the selection and at the bottom line there is an explanation of the command. ENTER invokes the respective command.

Stop: Quits the program execution and returns to menu.

Load file: Loads a mesh file from disk. The mesh was generated with the AUTOMESH program.

Plot Mesh: Plots the mesh currently in memory. The mesh can be anywhere in the screen.

Boundary: To specify boundary conditions. Its menu:

N finds the nearest node to the cross cursor. This cursor moves with the keypad direction keys.

The following two commands are related to the motion of the cursor cross:

F moves the cursor in larger steps (multiplication by 10)

S moves the cursor in smaller steps (division by 10)

E Exits to the main menu.

P specifies zero pressure at current node. Move cursor near the desired node and hit N. The cursor locks on the node. Hit P. A circle is placed around the node indicating zero pressure. For zero pressure on the boundary, answer Y to the question "Set Boundary to Po?" . Then the program asks for the number of nodes (partitions) in the peripheral-x and axial-y direction. Enter them. The nodes of the boundary are circled.

M when you specified all pressures on one arc, hit M to display the next arc to specify boundary conditions.

Analyze: The program performs the analysis and prints pressures at the nodes and bearing properties.

Post Proc: Plots the pressure distribution in a color code.

Make Model: To view and/or change default data. To keep default data just hit ENTER. To change data, enter new value and hit ENTER.

EXAMPLE: Find the eccentricity ration, attitude angle and the stiffness and damping coefficients of a 150° journal bearing with diameter d = 300 mm, length (width) L = 200 mm, radial clearance R/1000, viscosity η = 0.005 Pas, vertical load 22,000 N, one symmetric arc.

The input is prepared with <Make Model> and the input file BEARING1.DAT follows:

```
"Radius-omega-clearance-viscosity-eccentr-angle-load-noarcs"
.15,377,.00015,.005,4.713871E-05,56.24197,22000,1
"arc- angle1-angle2-width-preload"
1,-75,75,.2,0
```

The MESH8X7.DAT file was used to write the results. The file follows:

```
"Radius-omega-clearance-viscosity-eccentr-angle-load-noarcs"
.15,377,.00015,.005,4.713871E-05,56.24197,22000,1
"arc- angle1-angle2-width-preload"
1,-75,75,.2,0
"Sommerfeld number"," eccentricity ratio","length/D"," angle"
.818616,.314258,.6666666,56.24197
" kxx"," kxy"," kyx"," kyy"
3.879232E+08,5.84095E+08,5.106766E+07,8.497302E+07
" cxx"," cxy"," cyx"," cyy"
```

```
1724710,-126187.6,353487.9,123154.2
" SXxx, SXyy, SYxx, SYyy, SXxy, SYxy "
2.355746E+12,5.285694E+12,1.060735E+12,2.382544E+12,1.424733E+13,6.44432E+12
"SXyyy, SYyyy, SXxxx, SYxxx, SXxxy, SYxxy, SXxyy, SYxyy"
2.912916E+17,1.583138E+17,8.605323E+16,4.659823E+16,1.296481E+17,7.03883E+16,
1.933287E+17,1.048106E+17
```

S are higher order bearing coefficients. The screens produced during the execution follow:

PROBABLE ERRORS: Execution breaks: Very large mesh for the screen. Hardware incompatibility, wrong graphics adaptor. Out of memory. Absence of the COMPUME.SYS file. Execution breaks at post processing: Missing EGA card. Division by zero: data missing, you did not load mesh or boundary conditions.

```
BEARING DATA-hit ENTER for default data

Bearing Radius                              .15 ?            Exit MENU
Bearing Ang. Velocity      (rad/sec)        377 ?         > Load Bearg
Bearing Radial Clearance (Typical: R/1000)  .00015 ?        Load mesh
Oil Dynamic Viscosity (Default:SAE 5, 85oC) .005 ?          Plot Mesh
Bearing Eccentricity (guess, say Clear/2)   4.713871E-05 ?  Boundary
Attitute Angle,deg.(guess, say 45o)         56.24197 ?      Analyze
Bearing Vertical Load (Static)              22000 ?         Post Proc
Number of arcs  (maximum 10)                1 ?          > Bearg Data
                                                           Save Bearg

Arc no  1
Entrance Angle,deg. (From Lower point CCW)   -75 ?
```

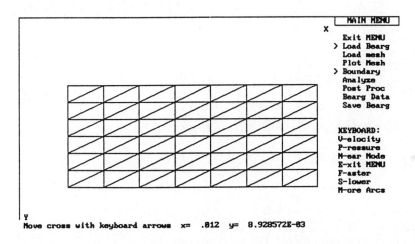

Move cross with keyboard arrows x= .012 y= 8.928572E-03

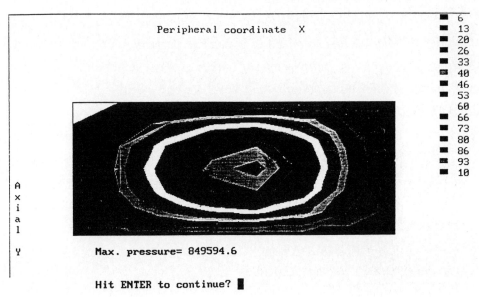

Max. pressure= 849594.6

Hit ENTER to continue? ▮

References and further reading

Baldwin, J. F. 1985."Fuzzy Sets and Expert Systems". Information Sciences, vol. 36, pp. 123-156.

Bishop, R.E.D., Johnson D, 1979, 2nd ed. *Mechanics of Vibration*. Cambridge: Cambridge University Press.

Bringham, E.O., 1974. *The Fast Fourier Transform*. Englewood Cliffs, N.J.: Prentice-Hall.

Clayton Labs (1990)."EXPERTS: An Adoptive Expert System Shell-User's Manual". St. Louis.

Clayton Labs (1988)."RODYNA: A General Purpose Rotor Dynamic Analysis Program". St. Louis.

Clough, R.W., Penzien, J., 1975. *Dynamics of Structures*. New York: Mc Graw-Hill Book Co.

Cooley, J.W., Tuckey, J.W., 1965. "An Algorithm for the Machine Calculation of Complex Fourier Series", J. Math. Comp., Vol. 19, No. 90, pp 297-301.

Dimarogonas, A.D., 1976. *Vibration Engineering*. Minneapolis: West Publishers.

Dimarogonas, A.D., Paipetis, S.A., 1983. *Analytical Methods in Rotor Dynamics*. London: Elsevier-Applied Science.

Dimarogonas, A.D., 1989. *Computer Aided Machine Design*. Englewood Cliffs, N.J.: Prentice Hall.

Dimarogonas, A.D., Haddad, S.D., 1992. *Vibration for Engineers*. Englewood Cliffs, N.J.: Prentice Hall.

Faddeeva, V.N., 1959. *Computational Methods of Linear Algebra*. New York: Dover Publications.

Federn, K., 1977. *Auswuchttechnik*. Berlin: Springer Verlag.

Gash, R., Pfuetzner, H., 1975. *Rotordynamik*. Berlin: Springer Verlag.

Gerald, C.F., Wheatley, P.O., 1984. *Applied Numerical Analysis*, 3rd Ed. Redding: Addison-Wesley Publishing Co.

Goodwin, M.J., 1989. *Dynamics of Rotor-Bearing Systems*. London: Unwin Hyman.

Juvinall, R.C., Marshek, K.M., 1991. *Fundamentals of Machine Component Design*. New York: J. Wiley.

Ka'rma'n, T. von, Biot, M.A., 1939. *Mathematical Methods in Engineering*. New York: Mc Graw-Hill Book Co.

Klotter, K., 1960. *Technische Schwingungslehre*. Berlin: Springer Verlag.

Lalane, M., Ferraris, G., 1990. *Rotordynamics Prediction in Engineering*. New York: J. Wiley and Sons.

Lazan, B.J., 1968. *Damping of Materials and Members in Structural Mechanics*. Oxford: Pergamon Press.

Meirovitch, L., 1980. *Computational Methods in Structural Dynamics*. The Netherlands: Sijthoff & Noordhoff.

Meirovitch, L., 1986. *Elements of Vibration Analysis*. New York: Mc Graw-Hill.

Moon, F, 1987. *Chaotic Vibration*. New York: John Wiley & Sons.

Nikravesh, P.E., 1988. *Computer-Aided Analysis of Mechanical Systems*. Englewood Cliffs, NJ: Prentice Hall.

Pestel, E, Leckie, F.A., 1963. *Matrix Methods in Elastomechanics*. New York: Mc Graw-Hill Book Co.

Poincare', H., 1892. *Sur les courbes de'finies par une e'quation differentielle*. Paris: Oevres, Gauthier-Villars.

Ramirez,R.W.., 1985. *The FFT Fundamentals and Concepts*. Englewood Cliffs, NJ: Prentice-Hall.

Randall, R.B., 1977. *Frequency Analysis*. Naerum, Denmark: Bruel & Kjaer.

Rayleigh, J.W.S., 1894. *Theory of Sound*. New York: Dover Publ. (1946).

Runge, C., 1904, *Theorie und Praxis der Reihen*. Leipzig.

Rosenblatt, F. (1962). *Principles of Neurodynamics*, Spartan, Washington, D. C.

Seireg, A., 1969, *Mechanical Systems Analysis*, Scrandon, Pa: International Textbook Co.

Shigley, J.E., 1990, *Mechanical Engineering Design.* New York: Mc Graw-Hill.

Shore, J.S., 1980. "Turbomachinery Problems and their Correction", in "Sawyer's Turbomachinery Maintainance Handbook", vol. 2, Chapter 7,, Connecticut.

Spectral Dynamics Corp., 1990. *Vibration Handbook.*

Stoker, J.J., 1950. *Nonlinear Vibrations.* New York: Interscience.

Thomson, W.T., 1978. *Vibration Theory and Applications.* Englewood Cliffs, N.J.: Prentice-Hall.

Timoshenko, S.P., Young, D.H., 1937, *Vibration Problems in Engineering.* 2nd Ed., New York: D. van Norstand Co.

Wilcox, J.B., 1967, *Dynamic Balancing of Rotating Machinery.* London: Pitman.

APPENDIX I
SYSTEMS OF UNITS

Symbol	Quantity	B.S. Units (lb-force)	Multiply by	to obtain SI Units
a	Acceleration	ft/sec^2	0.3048	m/sec^2
ω	Angular	rad/sec	1	rad/sec
	Velocity	(RPM	0.1047	rad/sec)
A	Area	ft^2	0.09290	m^2
I	Area Moment of Inertia	ft^4	86.30×10^{-4}	m^4
c	Damping Constant, linear	lbsec/ft	2101	Nsec/m
c_T	Damping Constant, Rotary	lbftsec/rad	0.1360	Nmsec/rad
E,V,T	Energy	ft-lb	1.356	J (Joule)
F	Force	lb	4.448	N
n	Frequency	sec^{-1}	1	sec^{-1}
L	Length	ft	0.3048	m
m	Mass	$lbsec^2ft^{-1}$	14.59	kg
J	Mass Moment of inertia	$inlbfsec^2$	.1130	kgm^2
M	Moment	ft-lb	1.356	Nm
P	Power	$lbftsec^{-1}$	1.356	W
		(hp	745.7	W)
P	Pressure	psi	6894	Pa (Pascal,N/m^2)
k	Stiffness, linear	lb/ft	14.59	N/m
k_T	Stiffness, Rotary	lbft/rad	0.009415	Nm/rad
t	Time	sec	1	sec
u	Velocity	$ft\ sec^{-1}$	0.3048	$msec^{-1}$
μ	Viscosity	$lbsecin^{-2}$	6894	Pasec
W	Work	lbft	1.356	J

144

APPENDIX II
ERROR CODES

Error code	Error
4	Out of data
5	Illegal function call or wrong monitor card
6	Overflow
7	Out of memory
9	Subscript out of range
11	Division by zero
25	Device fault
51	Internal error
52	Bad file name
53	File not found
54	Bad file mode
57	Device I/O error
61	Disk full
62	Input past end of file
64	Bad file name
71	Disk not ready
75	Path not found

END-USER LICENSE AGREEMENT

READ THIS AGREEMENT CAREFULLY BEFORE BUYING THE BOOK AND USING THE SOFTWARE:

The software included in the diskette(s) contained in the package on the opposite page ("Discs") is an educational version of MELAB and third party software given to the buyer of the book free of charge. Further distribution of third party software is subject to restrictions imposed by the respective copyright owners. The buyer of the book is hereby licensed to use the MELAB-educational version software under the following conditions:

Licence: You may use the Disks in any compatible computer, provided that the Disks are used on one computer and by one user at a time. You may capture screens and publish them without written permission, provided that the source will be properly cited.

Restrictions: You may not commercially distribute the programs in a manner that may infringe any copyright of the publisher, the author and third party software suppliers. You may not alter in any way the programs in the Disks. Integrating some of the software in larger packages is permitted, provided that it is for unsupported research and that full reference will be given to the author and publisher.

It is specifically stated that the software should not be used for commercial, institutional or governmental purposes.

Warranty: Beyond the publishers warranty for the physical disks been free of defects at the time of purchase, the programs are provided **"as is"** without any warranty of any kind, expressed or implied for the accuracy, the completeness or the results obtained from using the disks and the programs. Since the applicability, convergence and accuracy of numerical and approximate methods depend on the problem, it is the user's responsibility to establish the credibility of each program for the particular class of problems he or she is working with.

The programs included in the disks are for educational use only. Licence for commercial use, such as in product design, consulting, litigation, etc, is not hereby granted. If, however, a user applies the programs for the above purposes, it is his or her obligation to find if they are applicable. Moreover, the user must make sure that the programs comply with codes and standards for particular applications.

Code Variants: The size of code made necessary to compile the programs for size optimization, to be able to package them in a small number of floppy diskettes and reduce the book price. Users can obtain copies of the code in different compilations by sending to the author's laboratory, for the cost of compilation, postage and handling, a $60.00 check payable to Washington University and the specifications for the form of the desired code, which can include one or more of:

(a) 8086/8088 PC code, (b) compiled for speed optimization, (c) CTRL+BREAK key active, (d) uncompressed code, (e) 360/720/1200/1440 kb diskettes, (f) stand-alone files (without the BRT71ENR.EXE file)

to the address: **MELAB, Machine Design Laboratory, Campus Box 1185, Washington University, St. Louis, Mo 63130**.

The professional version of **MELAB** can be obtained from **Clayton Labs, 7135 Pershing Ave., St. Louis, Mo 63130**. Source code for course assignments or research, for instructors who have adopted the author's textbooks, can be obtained from the author.

146